T0324953

Cryogenic systems that involve two-phase (vapor–liquid) flows are widely used in industries such as aerospace, metallurgy, power engineering, and food production, as well as in high-energy physics research.

The purpose of this book is to describe characteristic features of cryogenic systems involving two-phase flow, create mathematical models of these systems, and then show how the models may be used to develop optimal designs for practical cryogenic systems. The models are examined using analytical and numerical techniques, and then the predictions are compared to experimental measurements. Since transient phenomena can produce severe and unexpected effects in cryogenic systems, the authors pay particular attention to this important topic.

Examples in the book are drawn from cryogenic fluid transport, gasification, and the stabilization of superconducting magnets. Much of this work is related to the development of large Russian systems in the areas of space technology, energy research, and particle physics.

This book, the first devoted solely to cryogenic two-phase flow, will be a valuable reference for cryogenic engineers and scientists.

CRYOGENIC TWO-PHASE FLOW

CRYOGENIC TWO-PHASE FLOW

Applications to large-scale systems

N. N. FILINA

NPO Cryogenmash
Balashihka, Russia

J. G. WEISEND II

Deutsches Elektronen-Sychrotron
DESY, Hamburg, Germany

CAMBRIDGE
UNIVERSITY PRESS

CAMBRIDGE
UNIVERSITY PRESS

32 Avenue of the Americas, New York NY 10013-2473, USA

Cambridge University Press is part of the University of Cambridge.

It furthers the University's mission by disseminating knowledge in the pursuit of
education, learning and research at the highest international levels of excellence.

www.cambridge.org
Information on this title: www.cambridge.org/9780521481922

© Cambridge University Press 1996

First published 1996
First paperback edition 2010

A catalogue record for this publication is available from the British Library

ISBN 978-0-521-48192-2 Hardback
ISBN 978-0-521-16840-3 Paperback

Contents

Preface

The subject of this book is cryogenic two-phase flow. Because of their very low boiling points, cryogenic fluids frequently exist in the two-phase state. In some cases this is intentional, such as when using the constant temperature of a two-phase flow to stabilize superconducting magnets. In other cases, the two-phase mixture is an undesirable consequence of heat leak or of a sudden change in the cryogenic system. An example of this is the creation of two-phase flow in a long cryogenic transfer line due to intrinsic heat leak.

Planned or not, the presence of two-phase flows in practical cryogenic systems is often a fact of life. The proper design of such systems requires an understanding of such flows and the ability to predict the effect of the two-phase flow on the system. This book describes techniques for modeling two-phase flows so that useful predictions can be made. The models are derived from the basic conservation laws. However, some empirical data is frequently required to complete the model. Once finished, the model is examined using analytical and numerical techniques. The predictions are then compared to experimental measurements.

The examples in this text are drawn from three important areas of cryogenic technology: gasification, fluid transport, and cryostabilization. These topics are all practical large-scale applications of cryogenic engineering. The methods described in this book can be applied to other cryogenic systems provided the underlying assumptions still hold true. Since transient phenomena can produce severe and unexpected effects, a significant portion of the text is spent discussing transient two-phase flow in cryogenic systems.

Chapter 1 gives a very brief introduction to cryogenic two-phase flow as well as discussing the problems associated with gasification, transport, and cryostabilization systems. Chapter 2 develops the conservation laws for cryogenic two-phase flows and uses them to model steady-state two-phase flow in a variety of large-scale systems. Chapter 3 introduces transient two-phase flow, stressing the relationship between the transient disturbance and the response time of the system. Chapter 4 models transient two-phase flow in gasification systems, and Chapter 5 looks at the transient behavior in magnet-stabilization systems.

The original manuscript of this book was written in Russian by N. N. Filina, based in large part on research she conducted. This was then translated into English by Valentina V. Savchenko of NPO Cryogenmash. J. G. Weisend II then completely revised the English text, checked the equations, and added additional material and references.

The authors wish to thank Professor William Schiesser of Lehigh University for his support and encouragement of this work. Dr. Philippe Lebrun of CERN and Professor Randall Barron of Louisiana Tech University both read a draft of the book and provided many useful comments. V. I. Datskov, Y. P. Filippov (both of JINR-Dubna), M. Levin (formerly of the SSC Laboratory), and S. Zinchenko (IHEP-Protvino) all helped to maintain contact between the authors. Florence Padgett, Physical Science Editor of Cambridge University Press, and her assistant Adam Tempkin are to be thanked for their assistance and enthusiasm.

N. N. Filina wishes to thank her teachers and colleagues at NPO Cryogenmash (I. M. Morkovkin, V. N. Krichtal, Y. E. Grakov, and M. I. Dyachuc) and the Institute of Mechanics at Moscow State University (R. I. Nigmatulin and V. E. Kroshilin). She is particularly grateful to her father, Professor Nikolai V. Filin, who has been a leader in Russian cryogenics for the past 25 years.

J. G. Weisend II wishes to thank his colleagues at the SSC Laboratory, particularly M. McAshan and J. Tompkins, and at the Centre d'Etudes Nucleaires Grenoble (G. Claudet and P. Seyfert) for supporting his work on this project. He is indebted to his wife Shari and children David and Rachel for their understanding and support while completing this work.

Nomenclature, SI units

English symbols

A = Area, m^2

C = Speed of sound, m/s

D = Channel diameter, m

d = Characteristic dimension, m

Fw = Specific frictional force between wall and two-phase flow, N/m^3

$F_{1,2}$ = Specific frictional force between phases, N/m^3

g = Gravitational acceleration, m/s^2

G = Mass flowrate, kg/s

I_i = Enthalpy of the ith phase, J/kg

$J_{1,2}$ = Liquid evaporation intensity (rate), kg/m (m^3/s)

k = Thermal conductivity, W/m^2 K

L = Channel length, m

P = Pressure, Pa

Q_w = Volumetric wall heat flux, W/m^3

$Q_{1,2}$ = Volumetric heat flux between phases, W/m^3

R = Gas constant, J/kg K

S = Slip

T = Temperature, K

t = Time, s

u_i = Internal energy of the ith phase, J

V = Volume, m^3

v_i = Velocity of the ith phase, m/s

$v_{1,2}$ = Longitudinal velocity component of fluid undergoing phase transition at the liquid–vapor interface, m/s

x = Mass vapor concentration (quality)

z = Spatial coordinate, m

Greek symbols

α_i = Volumetric concentration (void fraction) of the ith phase

ρ_i^0 = Density of the ith phase, kg/m^3

ρ_m = Mixture density, kg/m^3

ζ = Heat of vaporization, J/kg

μ_i = Viscosity of the ith phase, Pa s

τ = Time, s

τ^* = Time of disturbance propagation, s

Dimensionless parameters

Re_i = Reynolds number for the ith phase

Fr = Froude number

Nu = Nusselt number

Pr = Prandtl number

M = Martinelli number

Sh = Strouhal number

Eu = Euler number

Index

g = Gas

l = Liquid

o = Initial conditions

s = Equilibrium state parameters

v = Vapor

w = Wall

1 = Vapor

2 = Liquid

$2ph$ = Two-phase

1

Introduction to cryogenic systems with two-phase flows

1.1 Cryogenic gasification systems

Gasification systems are widely used in industry. Cryogens are transported and stored in liquid form, gasified, and delivered to the customer as a gas within given parameters. These systems typically consist of liquid storage tanks, transfer pipelines, and vaporizers in which the phase transition occurs.

An example of a gasification system is found at the Russian Baikonur launch complex. Here, liquid nitrogen at a pressure of 1–1.2 MPa and flowing at 13 kg/s is vaporized and used for a variety of purposes including purging and providing an inert atmosphere in spacecraft tanks.

Experience with gasification systems has shown that under transient conditions oscillations may develop. These oscillations are related to the properties of the cryogen, the design of the gasification system, and the nature of the disturbance. When the consuming equipment is connected or disconnected from the gasification system the product flowrate changes. The resulting disturbance propagates back to the liquid feed tank. If a large enough tank is used, the variation in the flowrate does not significantly affect the liquid parameters of the tank. In this case, the disturbance may be analyzed in terms of the subsystem consisting of the liquid and gas mains, vaporizers, and structural elements such as bellows and elbows.

Figures 1.1 and 1.2 show a commercial gasification system. The system supplies the gaseous product to the facilities of several consumers, each of which is connected by valves to the system. Liquid nitrogen from tanks (see Figure 1.3) is fed to the liquid main (0–1–2–3–4–5) and then into the evaporator (5–6). The resulting gas enters one of the consumer's facilities (8) via the gas main 6–7–8. The pressure at 0 is kept constant by a regulator. Table 1.1 lists the number, type, and hydraulic characteristics of the components within which the transient resulting from variation of the flow rate is localized (Dyachuk 1991).

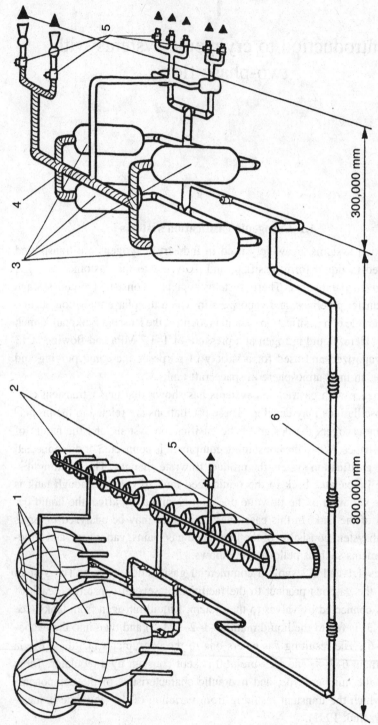

Figure 1.1. Example of a commercial gasification system: (1) liquid storage tanks, (2, 3) gasifiers, (4) gas mains, (5) outlet valves, (6) vacuum insulated liquid main (Dyachuk 1991).

800,000 mm

300,000 mm

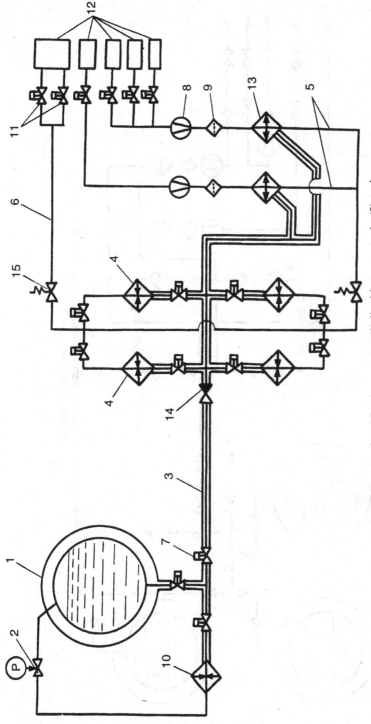

Figure 1.2. Schematic of gasification system: (1) liquid storage tank, (2) tank pressurization regulator, (3) liquid main, (4) gasifiers, (5, 6) gas main, (7) pneumatic valve, (8) flow meter, (9) filter, (10) tank pressurization evaporator, (11) outlet valve, (12) product consumer, (13) cooler, (14) check valve, (15) relief valve.

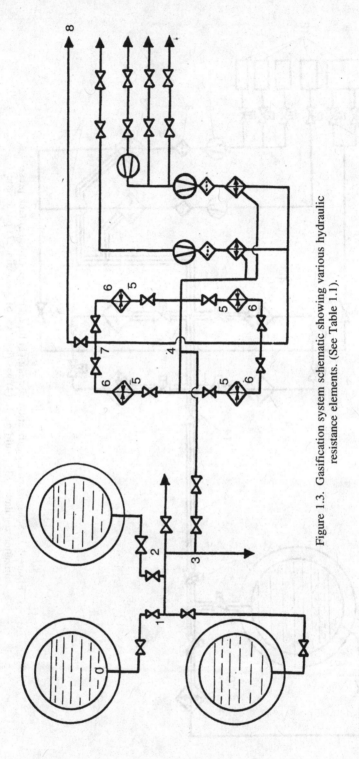

Figure 1.3. Gasification system schematic showing various hydraulic resistance elements. (See Table 1.1).

4

Table 1.1 *Description of elements in Russian IG-type gasifiers*

Description	Quantity	Hydraulic resistance coefficient (ξ)
Section 0–1, (liquid) dia. = 0.3 m		
Straight section dia. = 0.3 m	48 m	2.6
Turn 90 deg., dia. = 0.3 m	4	2.2
Hinged compensator	6	4.8
Valve	2	10.0
Tube inlet dia. = 0.3 m		1.0
		$\Sigma\xi(0–1) = 20.6$
Section 1–2 (liquid) dia. = 0.2 m		
Merging of flow	1	1.0
Straight section	1.5 m	0.1
Bellows dia. = 0.2 m	1	0.2
		$\Sigma\xi(1–2) = 1.3$
Section 2–3 (liquid) dia. = 0.2 m		
Flow separation dia. = 0.2 m	1	1.25
Straight section dia. = 0.2 m	3.5 m	0.3
Bellows dia. = 0.2 m		0.2
		$\Sigma\xi(2–3) = 1.75$
Section 3–4, (liquid) dia. = 0.2 m		
Straight section dia. = 0.2 m	785 m	58.9
Flow separation	1	1.25
Hinged compensator	3	2.4
Valve	2	10.0
Bellows	19	38
Turn, 90 deg.	8	4.3
Turn 135 deg.	1	0.4
		$\Sigma\xi(3–4) = 115.3$
Section 4–5 (liquid) dia. = 0.1 m		
Flow separation	1	1.25
Straight section	5.0 m	0.8
Bellows dia. = 0.1 m	1	0.6
Valve	1	5.0
Turn, 90 deg.	1	0.52
		$\Sigma\xi(4–5) = 8.2$
Section 6–7 (gas) dia. = 0.3 m		
Straight section dia. = 0.3 m	8 m	0.4
Turn, 90 deg.	3	1.6
Valve	1	5
		$\Sigma\xi(6–7) = 7$
Section 7–8, (gas) dia. = 0.3 m		
Straight section dia. = 0.3 m	300 m	15
Flow separation	1	1.25
Transition piece 0.4–0.3 m dia.	1	0.1
Valve	1	5
Turn, 90 deg.	10	5.4
		$\Sigma\xi(7–8) = 28$

As can be seen in the table, commercial gasification systems are complex, consisting of many different components including tanks, vaporizers, bellows, and valves. Transient oscillations in such systems cannot be described solely by analytical means (Filin & Bulanov 1985). Empirical results using standard measuring instruments (with an accuracy of 5 percent) are used to help understand the transient processes and to develop physical and mathematical models.

1.1.1 Cryogenic liquid gasifiers (vaporizers)

In gasification systems the phase transition is carried out in specialized apparatus known as vaporizers or gasifiers. In large-capacity vaporizers the gasification is caused by heat transfer from hot water or a superheated vapor to the cryogenic fluid. The performance of these devices can be improved by the augmentation of heat transfer between the two streams (Delhaye, Giot, & Riethumuller 1981; Stirikovich, Polonsky, & Tsiklauri 1982). In particular, use is made of different vortex generators (artificial roughness, spirals) to intensify the evaporation. The use of corrugated perforated packings with fins arranged in a staggered order (see Figure 1.4) has been shown to be one of the most efficient methods of heat-transfer augmentation.

Another type of vaporizer, known as a cold gasifier, uses heat exchange with the ambient environment to vaporize the cryogenic fluid. These gasifiers are characterized by a simple and reliable design and do not require additional power sources. Structurally, they consist of a set of parallel heat-exchange channels with developed surfaces. The free convection heat-transfer process on these surfaces is complicated by moisture condensation, desublimation of water and carbon dioxide, and in some cases by air condensation.

A proper understanding of two-phase cryogenic flow in vaporizers is necessary not only to optimize the heat transfer but also to predict the optimum flow pattern. The performance of vaporizers (gasifiers) of cryogenic fluids can be improved by controlling the hydrodynamics and heat transfer of the two-phase (vapor–liquid) flow. An example of this is seen in the operation of the IG type of vaporizers (see Figure 1.5) in the gasification system described in Table 1.1. During transient conditions in this system these gasifiers serve as generators of low-frequency oscillations (Nigmatulin, Filina, Kroshilin, & Dyachuk 1990). Analysis has suggested a number of modifications that include the creation of buffer volumes to reduce the pulsations (Dyachuk 1991). Figure 1.6 shows the cross sections of two different types of gasifier tubes.

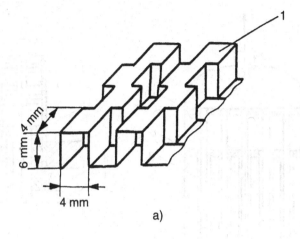

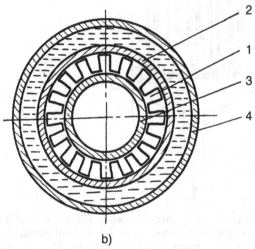

Figure 1.4. Elements of a gasifier channel: (a) corrugated fin insert. (b) Cross section of gasifier channel: (1) corrugated fin insert, (2) inner wall of heated pipeline, (3) hollow core, (4) outer wall of heated pipeline.

1.1.2 Liquefied natural gas (LNG) systems

A final example of the importance of gasification systems to the world economy is found in LNG systems. In Russia, large complexes have been built for the liquefaction, storage, and transport of LNG for commercial purposes. Natural gas is being used to provide alternative vehicle fuels and to provide energy for rural settlements. LNG gasification systems can be both fixed and

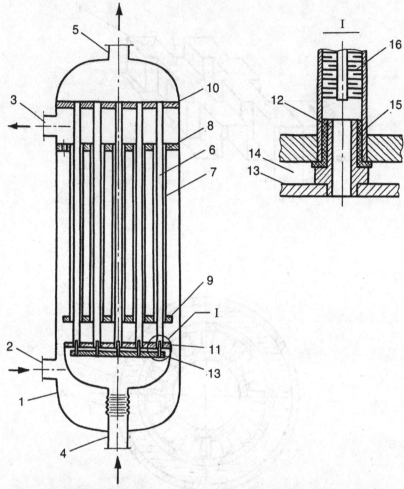

Figure 1.5. Type IG–30 gasifier: (1) casing, (2) inlet of heated fluid, (3) outlet of heated fluid, (4) inlet of cryogenic liquid, (5) outlet of produced gas, (6) inner tubes, (7) outer tubes, (8, 9) upper and lower tube sheets for outer tubes, (10, 11) upper and lower tube sheets for inner tubes, (12) branch pipe, (13) additional tube sheet, (14) clearance, (15) insulation, (16) corrugated fin insert (Dyachuk 1991).

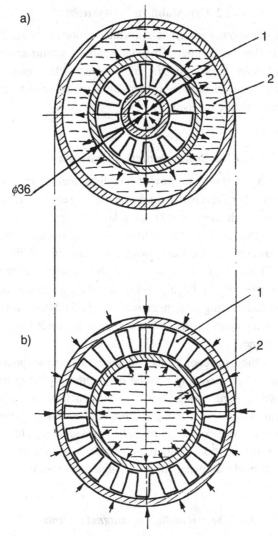

Figure 1.6. Cross sections of two different types of gasifier tubes: (1) cryogenic product space, (2) heated fluid. Arrows indicate temperature stresses.

mobile. The latter are found aboard various modes of transport – for example, ships, tractor trailer trucks, quarry dump trucks, and aircraft – that may be fueled by LNG. The problems associated with LNG gasification systems are the same as those found in other cryogenic gasification systems.

1.2 Cryostabilization systems

Cryostabilization is another application in which a good understanding of two-phase cryogenic flows is required. Cryostabilization systems are designed to maintain the temperature of cryogenic components within desired operating parameters. Two examples of these systems are large space-simulation chambers and superconducting magnets.

1.2.1 Space-simulation chambers

These chambers are designed to provide a realistic thermal test environment for spacecraft and other flight hardware. Since full-size equipment is frequently tested the size of the chambers can be quite large. In space-simulation chambers the cold of space is simulated by heat-absorbing shields cooled to cryogenic temperatures constructed from finned channels (Filin & Bulanov 1985; Gorbachev 1987). The heat load to these shields is determined by the test requirements and may vary in magnitude and duration. Stringent requirements are specified for maintaining the temperature of the shields within a wide range of heat loads as increases in the shield temperature result in degraded test chamber performance.

A common method for cooling the shields is to use two-phase LN_2. As shown in Figure 1.7, the cooling may be done either by using natural convection loops with boiling taking place in the channels or in a separate bath, or else by making the channels a stagnant boiling bath that is periodically refilled. Efficiency in these devices is related to providing a stable flow of the LN_2 with a lack of pulsations. This requires calculating the two-phase flow in the vertical channels, allowing for the changing hydrostatic pressure.

1.2.2 Superconducting magnet systems

In order to function properly, superconducting magnets must be maintained at a temperature low enough to avoid reverting to a normal electrically resistive state. Large experimental facilities developed for the study of plasma physics and high-energy physics have made extensive use of the high magnetic fields developed by superconducting magnets. In many cases the magnets are completely or partially cooled by two-phase flows.

The Russian plasma physics experiment Tokamak 7 uses superconducting coils to form part of its plasma containment system. Forty-eight copper stabilized niobium–titanium coils are contained in a toroidal cryostat with an

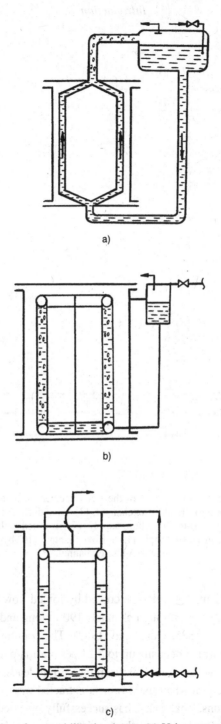

Figure 1.7. Examples of cryostabilization loops: (a) Using natural convection and a separate boiling bath. (b) Using natural convection and boiling in the channels. (c) Saturated bath cooling with liquid–vapor interface in the channels.

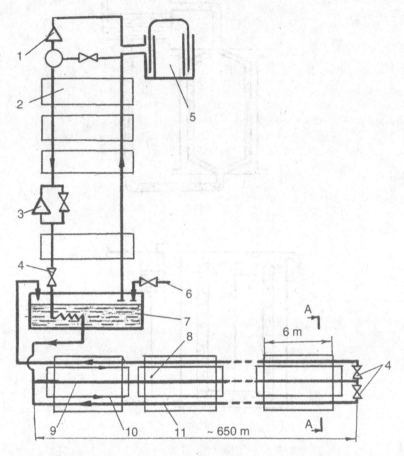

Figure 1.8. Schematic of one arm of the UNK accelerator magnet cooling system: (1) compressor, (2) heat exchanger, (3) expander, (4) throttle valve, (5) gas storage, (6) liquid helium supply, (7) saturated bath, (8) super-conducting magnets, (9) single-phase flow channel, (10) bypass channel, (11) two-phase channel.

outer radius of 1.22 m. The coils are cooled by forced flow two-phase or supercritical helium moving through channels 190 m long and 2 mm in diameter (Filin & Bulanov 1985; Gorbachev 1987). The cooling system removes eddy current heat (which deposits up to 5 kJ per plasma pulse at a rate of 30 pulses per hour) as well as heat leak from the current leads. Up to 300 kJ are deposited in the helium when the power is extracted from the magnet under emergency conditions. Tokamak 7 has successfully operated for hundreds of hours, providing valuable plasma physics data.

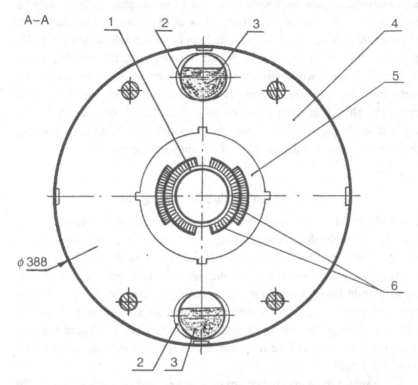

Figure 1.9. Cross section of UNK superconducting magnet: (1) single-phase flow channel, (2) bypass channel, (3) two-phase channel, (4) iron yoke, (5) collars, (6) superconducting cable.

Particle accelerators designed for the study of high-energy physics also make use of large superconducting magnet systems. Modern particle accelerators are among the largest scientific instruments ever constructed. Examples of super-conducting accelerators cooled principally by two-phase flow are the Tevatron at Fermilab (USA), the German HERA facility, and the Russian UNK-3000. The giant Large Hadron Collider (27 km in circumference) being constructed in Geneva will use a two-phase superfluid helium cooling loop. Studies in support of this unique cooling scheme are currently underway.

The UNK machine currently under construction south of Moscow will be 21 km in circumference and is divided up into 24 sections known as arms. Each arm contains about 110 superconducting magnets and is approximately 650 m in length. Figure 1.8 is a schematic of one of these arms while Figure 1.9 is a cross section of the helium space of one of the dipole magnets. The superconducting coils are cooled by a flow of single-phase helium, which is itself continuously cooled by a counterflow of two-phase helium. At the end of each 650 m arm, the single-phase helium is throttled through a valve. The resultant pressure drop causes the helium to change into a lower temperature two-phase mixture which then becomes the counterflowing stream that absorbs heat from the single-phase helium surrounding the superconducting coils. The physics of this method of magnet cooling is complicated, involving both single-phase and two-phase flows, counterflow heat exchangers, and the periodic mixing and separating of the single-phase stream.

1.3 Cryogenic fluid transport systems

The widespread use of liquid cryogens in commercial and scientific applications has led to the development of cryogen transport systems consisting primarily of large pipelines. The systems are designed to provide a cryogen at a given flowrate, pressure, temperature, and quality (vapor content). Movement of fluid through these pipelines can be achieved by the use of cryogenic pumps or via a pressure differential between the storage tank and the end user. Through the use of thermally insulated pipes and liquid subcooling, liquid cryogens can be transported through pipelines that extend kilometers in length.

As a result of the inevitable environmental heat leak, the temperature of the cryogenic fluid increases as it flows through the pipeline. When the fluid temperature reaches the saturation temperature corresponding to the local pressure, boiling occurs. The resulting two-phase flow is undesirable and a prediction of the cryogen flow parameters requires an understanding of the possible vapor content of the fluid.

Cryogenic transport systems with large vertical sections (such as the liquid oxygen transport system designed in support of space technology in Russia) present unique problems. In addition to the difficulties associated with horizontal systems, the vertical systems are complicated by the effect of the hydrostatic head on vapor generation and the possible emergence of the geyser effect (Filin & Bulanov 1985). This last effect may occur at very low velocities in the pipeline or when the flow is suddenly stopped. During geysering there is a rapid boiling of the cryogen followed by the ejection of liquid from the vertical pipeline. If the supply tank is located above the vertical section it will backfill and be subject to an impact of "water hammer."

The boiling and impact filling are cyclical in nature and can result in the destruction of the transport system. The possibility of geysering must be considered when designing transport systems with vertical sections, and precautions should be taken to prevent it in those systems. Solving this problem also requires the ability to model two-phase flows.

1.4 Two-phase (vapor–liquid) flows

The vapor–liquid flows addressed in this work are heterogeneous (two-phase) mixtures. In heterogeneous mixtures the inclusions or inhomogeneities are macroscopic in scale. This may be contrasted to homogeneous mixtures, in which the components are mixed on a molecular level.

When modeling the processes and motion of a heterogeneous mixture the assumption is made (Nigmatulin 1990) that the characteristic dimensions of the inclusions and inhomogeneities in the mixture (e.g., the diameter of droplets and bubbles, film thickness) are much larger than the molecular–kinetic distance. This means that the inclusions contain a large quantity of molecules, which allows the use of classical continuum mechanics of a homogeneous media for describing the processes associated with the inhomogeneities. Physical properties (for example, viscosity and thermal conductivity) of each of the two phases are described in the framework of the one-phase state.

Another assumption that is made in the modeling of the two-phase systems discussed here is that the dimensions of the inhomogeneities and the channel diameter are much much less than the channel length, over which the phase parameters may vary significantly. Thus, the problem being studied is one-dimensional. For the large commercial systems addressed in this work this assumption is valid and applies to both steady-state and transient flow conditions.

It is worthwhile to define several terms that are used to describe two-phase flow. The first term is the void fraction (α_1). This is defined as the ratio between the volume of the vapor and the total volume of the flow. Since many practical two-phase flow problems are one-dimensional it is common to describe the void fraction in terms of a ratio of areas. Thus:

$$\alpha_1 = \frac{A_v}{A_v + A_l},$$ (1.1)

where

A_v is the area occupied by vapor at a given point in the system,
A_l is the area occupied by liquid at the same point.

Various methods are used to measure the void fraction. These methods include the method of dynamic scales (measuring the pressure on a disk normal to the flow), attenuation of α and γ radiation, photographic techniques, and techniques using microthermocouples, lasers, and NMR devices (Delhaye et al. 1981). In addition, capacitive methods have been extensively used in the measurement of void fractions in cryogenic two-phase flow (Filippov & Alexeyev 1994; Hagedorn, Leroy, Dullenkopf, & Haas 1986; Katheder & Sußer 1991).

Similar, though distinct from the void fraction, is the quality (x). In a nonflowing two-phase system, this is defined as the ratio of the vapor mass to the total mass. In a flowing system it is the ratio of the liquid and vapor mass flowrates:

$$x = \frac{\dot{m}_v}{\dot{m}_v + \dot{m}_l}. \tag{1.2}$$

In a nonflowing system, α_1 and x are related to each other only by the liquid and vapor densities. However, this is not necessarily true in all two-phase flows. The reason is that in two-phase flows the velocities of the phases are frequently different. The vapor phase tends to move at a higher speed than the liquid phase. The ratio of these velocities is known as the slip (S) and is given by

$$S = \frac{v_v}{v_l}. \tag{1.3}$$

From equations (1.1)–(1.3) and the definition that $\dot{m} = \rho v A$ it can be shown that

$$S = \left(\frac{x}{1-x}\right)\left(\frac{1-\alpha_1}{\alpha_1}\right)\left(\frac{\rho_l}{\rho_v}\right). \tag{1.4}$$

It is generally not possible in two-phase flows to determine the void fraction directly from the quality; the slip between the phases must also be known. It is possible for a given set of fluid conditions to create plots of void fraction as a function of quality and various slip values. An example of such a plot is given in Figure 1.10 for saturated helium at 4.2 K.

1.4.1 Two-phase flow regimes

The various flow regimes or patterns that arise in two-phase flow significantly affect the hydrodynamic and heat-transfer behavior of the flow. These regimes are determined by a variety of factors such as the void fraction, density, viscosity, velocity, surface tension, channel geometry, and external heat load.

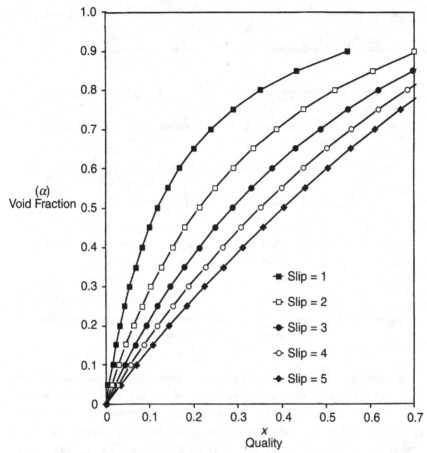

Figure 1.10. Calculated void fraction as a function of quality and slip for saturated helium at 4.2 K.

Figure 1.11 (Theilacker & Rode 1988) is an illustration of typical regimes in horizontal flow.

Although it is not possible to predict the presence of a regime based solely on the void fraction, the flow regimes change as the volumetric concentration of the vapor increases. Bubble flow occurs at values of $\alpha_1 < 0.2$–0.3. As the void fraction increases, the bubbles start to coalesce and slug flow is set up (0.2–$0.3 \le \alpha_1 < 0.6$–0.8). At $\alpha_1 \ge 0.6$–0.8, annular flow conditions are set up in which the liquid phase is confined to a film coating the walls of the channel. At even higher void fractions, droplet flow occurs in which the liquid phase exists only as droplets entrained in a vapor flow.

The transition between these regimes is diffuse rather than distinct and un-

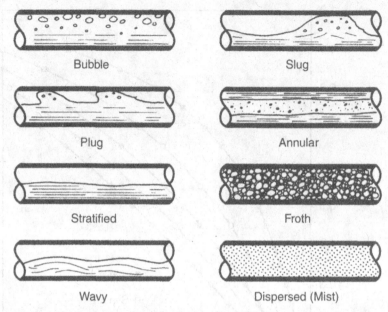

Figure 1.11. Examples of various flow regimes in horizontal two-phase flow
(Theilacker & Rode 1988).

stable conditions (emulsion, froth) as well as variations to these regimes also
exist. Gravitational forces play a significant role in this problem in both horizontal and inclined channels. Additionally, the transitions between various
regimes are dependent on the type of fluid under study. For example, the transitions derived from water–air studies are not directly applicable to two-phase
helium flows.

1.4.2 Models of two-phase flows

The two principal models that are used to describe two-phase flows are the
single-component model (also known as the homogeneous model) and the two-
component model. In the single component model, the two-phase mixture is
replaced by a single-phase medium. The physical properties of this medium,
such as density, viscosity, velocity, and temperature, are averages of the properties of each of the two phases, which can be defined in several ways. Additional modifications are made based on factors such as drift velocity, void fraction, and coefficient of slippage. Significant work has been done with this model
by Zoober, Findley, and Wallace (Delhaye et al. 1981). Unfortunately, single-
component models tend to be cumbersome and have limited application.

More successful is the two-component model in which the conservation laws (mass, energy, and momentum) are written for each phase separately with allowance for the interaction between the phases (Nigmatulin 1990). The interactions between the phases are defined by a distribution of parameters (stresses, temperatures, concentrations, etc.) around the inhomogeneities. In order to quantify these effects, the physical picture of the problem is generally greatly simplified. This simplification, along with nondimensional analysis, experimental results, and analogies from well-known single-phase cases, allows the development of models for the interactions between the phases. This in turn permits the calculation of the friction between the wall and the flow, the interfacial friction, and the heat transfer between the phases and between the wall and the flow. Significant work has been done in this area (Delhaye et al. 1981; Nigmatulin 1990).

1.5 Requirements for the design of practical systems with two-phase flows

When designing practical systems that use two-phase (vapor–liquid) flows, a variety of requirements must be met. The desired performance characteristics (temperature, pressure, flow rate, and void fraction) of the system must be achieved. The system should be easy to build and operate. Operation of the system should be as energy efficient as possible. High reliability is also necessary with the system having good fatigue, strength, and corrosion-resistance characteristics. Safety devices such as relief valves and emergency collectors should be designed into the system from the beginning to protect personnel and equipment.

An understanding of system behavior during nonstandard operating conditions is required so that the designer is sure that such conditions do not cause additional problems. For instance, the violent boiling that can occur when a superconducting magnet stabilized by two-phase helium reverts to the normally conducting state (a phenomenon called quenching) should not lead to system failure.

In order to optimize the design to meet these varied requirements, the ability to model and predict the physical behavior of two-phase flows is absolutely necessary.

2

Hydrodynamics and heat transfer in two-phase flows in cryogenic media

2.1 Physical features of cryogenic vapor–liquid flows

The basis for understanding two-phase (liquid–vapor) flows and applying that understanding to solve real-world problems is the development of physical models of the flows. To be complete, these models must describe the flow regime, physical parameters of each phase (e.g., pressure, temperature, and velocity), and the processes that occur at the phase boundaries and at the boundary between the fluid and the wall.

The most important parameter of two-phase flow is the structure or flow regime, because this defines the hydrodynamics and heat transfer of the problem. In the commercial cryogenic systems studied, any of the two-phase flow regimes (annular, bubble, slug, etc.) may develop. The type of flow structures seen in a system depend upon a variety of parameters including type of fluid, channel geometry (length, diameter), channel orientation, external heat input, thermophysical properties of the phases, pressure and temperature of the phases, and the quality of the heat-transfer surface.

Direct measurement of the flow regime is difficult because of transient effects and interaction of the measuring probe with the flow and may require sophisticated techniques such as gamma ray or neutron scattering. Thus, the use of theoretical models to describe the flow regime is very important. These models should be accurate, physically meaningful, and sufficiently simple. This is particularly important when describing the complicated commercial systems that involve cryogenic two-phase flow.

Models describing cryogenic two-phase flows typically benefit from both theoretical and experimental studies. An example problem is shown in Figure 2.1. Here a cryogenic fluid (LN$_2$) is entering a room temperature (300 K) pipeline. Since the wall temperature is far above the metastable superheat (Leidenfrost) temperature for the fluid, the liquid will be separated from the wall by its vapor. At the liquid–vapor interface the liquid and vapor temperatures

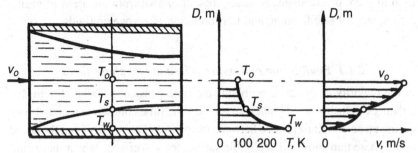

Figure 2.1. Cryogenic liquid flow in a warm pipeline: T_s = saturation temperature; T_w = wall temperature; T_o = inlet temperature; v_o = inlet velocity.

are equal and correspond to the saturation temperature (T_s) that is associated with the fluid pressure (P). In the vapor space the temperature gradient is given by:

$$\Delta T = T_w - T_s(P). \tag{2.0}$$

In this example, assume the LN_2 is subcooled with $T_o = 77$ K and $P_o = 101.3$ kPa. This yields T_s (P) = 104 K and results in the temperature gradient in the vapor space being $\Delta T = 300$ K $-$ 104 K = 196 K.

Since the LN_2 enters subcooled, the temperature distribution in the liquid jet will be different, ranging from initial inlet temperature (T_o) in the flow core to the saturation temperature (T_s) at the interface. With a knowledge of the characteristic temperatures at definite points, it is possible to calculate the parameter distribution at each point of the flow cross section.

The physics of the problem also shows that the phase velocities are different. As a result of the retardation or "sticking" of the flow at the wall, it is obvious that $v = 0$ at the wall. At the liquid–vapor interface, the tangential stresses are equal. Thus at the interface,

$$\tau = \mu_1 \partial v_1 / \partial z = \mu_2 \partial v_2 / \partial z. \tag{2.1}$$

That is, the ratio of the velocity gradients is proportional to the viscosity of the phases. (Note that in the course of this text subscript 1 generally refers to the vapor phase and subscript 2 generally refers to the liquid phase.)

In the remainder of this chapter, the general conservation equations of mass, momentum, and energy are developed for two-phase flows. These equations, along with appropriate simplifications and the use of empirical results, are then used to solve practical steady-state problems drawn from the areas of gasification, magnet stabilization, and transportation of cryogenic fluids.

2.1.1 Equilibrium and nonequilibrium two-phase flows

In the majority of practical problems the time constant associated with disturbances to the system is much larger than the time constant associated with setting up an equilibrium state between the phases. This results from the speed of the phase transition and negligible surface tension effects. Under these conditions the pressures in the phase may be assumed to be equal. Thus,

$$P_1(\rho_1^0, T_1) = P_2(\rho_2^0, T_2) = P. \tag{2.2}$$

This assumption may not be true in the case of explosive or other high-speed disturbances. In these cases, the dynamics of the interface processes and the phase pulsations will dominate the problem.

We can further divide two-phase flows into equilibrium and nonequilibrium flows. The distinction is very straightforward.

In *equilibrium* two-phase flows, the temperatures, pressures, and velocities of the phases are equal. Thus,

$$T_1 = T_2 = T_s(P); \qquad P_1 = P_2 = P; \qquad v_1 = v_2. \tag{2.3}$$

A typical example of an equilibrium flow is horizontal bubble flow. Here the phases are well dispersed and they have equal temperature, pressure, and velocity.

In *nonequilibrium* two-phase flows, the phases have different velocities and/or temperatures, resulting in a complicated distribution of parameters across the channel cross section. Many practical problems involve nonequilibrium flows.

It is important to keep in mind that in this context the terms equilibrium and nonequilibrium do not refer to the time dependence of the problem. Instead, the term "steady-state" will describe problems whose parameters are time independent, while "transient" will refer to those problems whose parameters vary with time.

The modeling of nonequilibrium flows as equilibrium flows can be a powerful analysis technique. In this method, average temperatures and velocities are assigned to both phases. While this technique greatly simplifies the physical picture, it does permit description of the differing parameters (e.g., compressibility, viscosity, and density) of the phases.

2.2 Conservation equations for heterogeneous two-phase flows

As discussed in Chapter 1, the subject of this text is heterogeneous two-phase flows that are described by classical continuum mechanics. The phases are assumed to be completely interpenetrating and the multiple-component model is used. This requires that for a unit volume, the conservation equations of mass, momentum, and energy be written for each component present, taking into account the phase transition, interaction between phases, and the interaction of the fluid and the wall.

The conservation equations of a heterogeneous nonequilibrium flow with N components are given by Nigmatulin (1990):

Mass

$$(\partial \rho_i / \partial t) + \nabla(\rho_i \, \vec{v}_i) = \sum_{j=1}^{N} J_{ji}, \qquad (2.4a)$$

Momentum

$$\rho_i(D\vec{v}_i/Dt) = -\alpha_i \nabla P + \rho_i \vec{g}_i$$

$$+ \sum_{j=1}^{N} [\vec{F}_{ji} + J_{ji}(\vec{v}_{ji} - \vec{v}_i)] + \vec{F}_{wi}, \qquad (2.4b)$$

Energy

$$\rho_i^0 \frac{Du_i}{Dt} = \frac{\alpha_i P D \rho_i^0}{\rho_i^0 \, Dt} + \sum_{j=1}^{N} \left[\begin{array}{c} \vec{F}_{ji}(\vec{v}_{ji} - \vec{v}_i) + J_{ji} \dfrac{(\vec{v}_{ji} - \vec{v}_i)^2}{2} + \\ J_{ji}(u_{ji} - u_i) + Q_{ji} \end{array} \right] - \nabla Q_i, \qquad (2.4c)$$

where

$i,j = 1, 2, 3, \ldots N; \quad i \neq j.$

$P_1(\rho_1^0, T_1) = P_2(\rho_2^0, T_2) = \ldots P_N(\rho_N^0, T_N) = P.$

$\rho_i =$ the component density.

$\alpha_i =$ the volumetric concentration (void fraction) of the ith component.

$$\sum_{i=1}^{N} \alpha_i = 1.$$

$$\rho_i^0 = \frac{\rho_i}{\alpha_i}.$$

$u_i = u_i(\rho_i^0, T_i)$ is the internal energy.

$J_{ij} = -J_{ji}$ is the intensity of the mass transition from component i to j per unit time and per unit volume.

$\vec{F}_{ij} = -\vec{F}_{ji}$ is the friction force between components.

Q_{ij} is the heat transferred between components.

$\vec{v}_{ij} = \vec{v}_{ji}$ is the longitudinal velocity of the substance undergoing phase transition at the interface.

$\vec{g}_i = gh_i$ is the gravitational effect on the ith component.

$\overrightarrow{F_{wi}}$ = the wall friction force on the ith component.

The value of N is determined by the number of structural components. In the case of bubble flow in a channel $N = 2$ (liquid and bubbles). If there is a vapor film at the wall and vapor-droplet flow in the center then $N = 3$ (film, droplets, and vapor). Taking the common case of two-component flows and assuming that the parameters are a function of only one spatial dimension (z), Equation (2.4) reduces to

Mass

$$\frac{\partial(\rho_i^0 \alpha_i)}{\partial t} + \frac{\partial(\rho_i^0 \alpha_i v_i)}{\partial z} = (-1)^{i+1} J_{2\,1}, \qquad (2.5a)$$

Momentum

$$\frac{\partial(\rho_i^0 \alpha_i v_i)}{\partial t} + \frac{\partial(\rho_i^0 \alpha_i v_i^2)}{\partial z} = -\frac{\partial(\alpha_i P)}{\partial z} +$$

$$F_{gi} + (-1)^{i+1}[F_{1,2} + J_{1,2}(v_{1,2} - v_i)] + F_{wi}, \qquad (2.5b)$$

Energy

$$\frac{\partial\left[\rho_i^0 \alpha_i \left(I_i + \frac{v_i^2}{2}\right)\right]}{\partial t} + \frac{\partial\left[\rho_i^0 \alpha_i v_i \left(I_i + \frac{v_i^2}{2}\right)\right]}{\partial z} +$$

$$\frac{\partial(\rho_i^0 \alpha_i P)}{\partial z} = F_{gi} v_i + (-1)^{i+1}\left[Q_{ig} + J_{2,1}\frac{(v_{1,2} - v_i)^2}{2}\right] + Q_{iw}, \qquad (2.5c)$$

where

F_{gi} = the gravitational force.

I_i = the enthalpy of the ith component.

Q_{iw} = the heat transferred from the channel wall into the ith component.

If the flow being examined is an equilibrium flow with small pressure drops in the channel, then all the external heat transferred into the flow goes into the phase transition and thus:

$$J^0_{2\,1} = Q_w/\zeta, \tag{2.6}$$

Where ζ is the heat of vaporization. However, If the vapor is superheated then it accumulates part of the external heat. Therefore,

$$J_{2,1} < J^0_{2,1}; \qquad T_1 > T_2 = T_s(P). \tag{2.7}$$

Equations (2.4) and (2.5) are the most general forms of the conservation laws for two-phase flows. In order to solve practical problems these equations must be simplified to describe the situation under study and empirical results for some of the terms in the equations must be included.

2.3 Gasifier channels with intensifiers

In cryogenic gasification systems, the gasifier serves as a heat exchanger in which external heat is applied to a cryogenic liquid or two-phase flow to convert that flow into a gas at a desired temperature. If the surface area of the phase transition within the gasifier channels is sufficiently developed, then the phases are well mixed, the average velocities of the two phases are equal, and all the external heat goes into evaporating the liquid. In this case, the temperatures of the phases are equal to each other and to the saturation temperature as determined by the flow pressure.

For example, in droplet flows, the surface area of the phase transition may exceed the surface area of the heated channel by a factor of 10^2–10^3. Under these conditions the droplets remove all the external heat. If, however, the surface areas of the phase transition and heated channel are comparable, then the external heat goes to both evaporating the liquid and superheating the vapor. This reduces the amount of heat removed from the channel wall. Thus, the most efficient gasifiers are designed to generate a two-phase flow regime that boosts the amount of external heat removed from the wall. In this case, dispersed flows with uniformly mixed phases are the optimum choice.

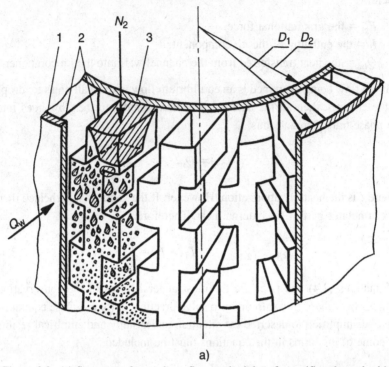

a)

Figure 2.2. (a) Structure of two-phase flow at the inlet of a gasifier channel with packing: (1) externally heated wall, (2) packing located in the annular space, (3) hollow core.

Artificial roughness and vortex generators (Delhaye et al. 1981) are among the techniques used to intensify the heat exchange between the channel wall and the flow. A type of intensifier that is particularly efficient in cryogenic gasifiers is the packing of the channel with fins arranged in a staggered order. This type of intensifier is shown in Figure 2.2. The liquid product enters the cellular structure with a heated outer wall. A jet flow with a complicated distribution of temperatures and velocities is generated in the first layer of cells. The jet impinges on the staggered fins in the second layer of cells and is crushed, resulting in a two-phase flow with separate jets, droplets, and superheated vapor mixed together. The close packing of the fins causes significant crushing of the cryogenic flow, which ultimately results in the desired droplet flow. Thus, the droplet flow comes about due to flow crushing rather than due to heat transfer from the wall as in the case of gasifiers with smooth tubes. Once droplet flow has been established, the temperatures of the liquid droplets and the vapor are equal to each other and to the saturation temperature associated with the local pressure. There will be some superheated vapor

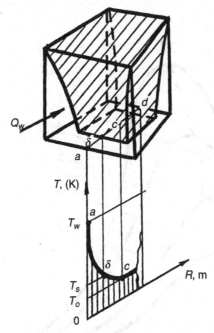

Figure 2.2. (b) radial temperature distribution in the first layer of packed cells.

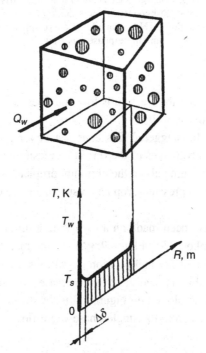

Figure 2.2. (c) radial temperature distribution in the equilibrium droplet flow.

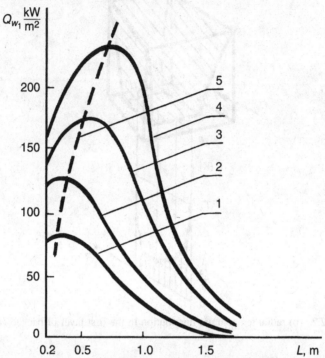

Figure 2.3. Wall heat flux along gasifier channel length at different mass flowrates (Kuzmenko 1979): (1) $\rho v = 35$ kg/sm^2; (2) $\rho v = 53$ kg/sm^2; (3) $\rho v = 83$ kg/sm^2; (4) $\rho v = 150$ kg/sm^2; (5) author's (N.F.) calculated curve separating two-phase flow from single-phase gas flow.

at the channel walls, but the fraction is low and does not significantly affect the heat transfer from the walls.

The presence of the staggered fins causes a larger pressure drop in the flow. However, the advantages of the fins in increasing the heat transfer surface and in rapidly bringing about the optimum droplet flow regime typically outweigh the increased pressure drop and result in a more efficient cryogenic gasifier.

Measurements have been made on a cryogenic gasifier with staggered fin intensifiers (Kuzmenko 1979). Figure 2.3 shows the measured wall heat flux (kW/m^2) as a function of channel position for various mass flowrates per unit area (kg/sm^2). Note that the wall heat flux increases to a maximum and then drops off sharply. Also plotted on Figure 2.3 is the theoretical curve separating the two-phase flow from the single-phase vapor flow. This curve was calculated assuming a steady-state liquid flow and allowing for the measured wall heat flux, the measured channel outlet temperature, and the variation of fluid parameters with temperature.

The peak in the measured wall heat flux and its relationship to the calculated two-phase vapor boundary has a straightforward physical explanation. As the two-phase mixture flows through the gasifier at constant mass flowrate, it absorbs heat from the channel wall resulting in an increase of the vapor fraction of the mixture. This reduces the average density of the mixture, which in turn increases the average velocity (since $\rho v A = $ constant). The higher average velocity increases the convective heat transfer at the wall, resulting in an increasing wall heat flux. As previously stated, the maximum wall heat flux occurs in the equilibrium droplet flow regime, in which all the heat from the wall goes into the droplets. Once the droplets are all evaporated, the wall heat goes into superheating the vapor. This decreases the temperature difference between the wall and the vapor, resulting in a decrease in the wall heat flux. Thus, the wall heat flux drops sharply after the fluid is converted from two-phase to single-phase vapor.

At higher flowrates, some of the droplets enter the superheated vapor region of the channel, resulting in a nonequilibrium two-phase flow. This effect is seen in the higher flowrate curves in Figure 2.4, which no longer have a monotonic behavior.

2.3.1 Mathematical description

In order to describe the behavior of cryogenic gasifiers with staggered fins in any detail it is necessary to solve the basic conservation laws (Equation 2.5). If we model the portion of the gasifier that contains equilibrium flow ($v_1 = v_2 = v$, $T_1 = T_2 = T_s(P)$) and assume the problem to be steady state, then the continuity and momentum equations may be written as

$$\partial(\rho_1^0 \alpha_1 v)/\partial z = J_{2,1}, \tag{2.8a}$$

$$\partial(\rho_2^0 \alpha_2 v)/\partial z = -J_{2,1}, \tag{2.8b}$$

$$\partial[(\rho_1^0 \alpha_1 + \rho_2^0 \alpha_2)v^2]/\partial z = -\partial P/\partial z - F_w. \tag{2.8c}$$

Thus, the governing equations are greatly simplified in the case of equilibrium flows. The next step is to start assigning values to the unknowns in Equation (2.8).

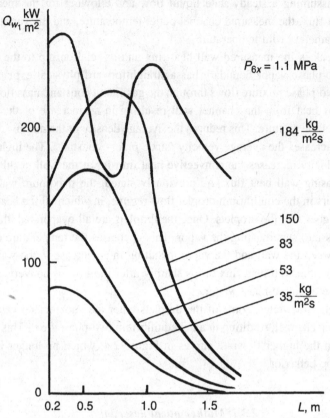

Figure 2.4. (a) Influence of increasing mass flowrate on gasifier channel wall heat flux at different pressures (Kuzmenko 1979).

One way to express the density of the phases is to assume the vapor is an ideal gas and the liquid is incompressible, in which case:

$$\rho_1^0 = P/RT_s(P), \qquad (2.9a)$$

$$\rho_2^0 = \text{constant}. \qquad (2.9b)$$

Alternatively both phases may be treated as compressible:

$$\rho_1^0 = (P(T_s)/Z_1(T_s)RT_s) \qquad (2.10a)$$

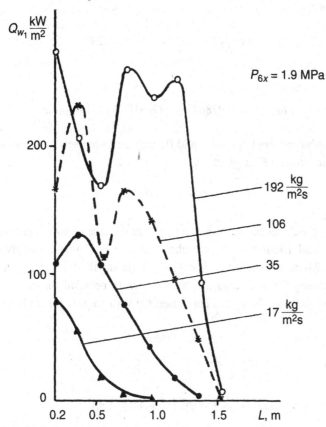

Figure 2.4. (b) Influence of increasing mass flowrate on gasifier channel wall heat flux at different pressures (Kuzmenko 1979).

and

$$\rho_2^0 = Z_2(T_s)T_s, \tag{2.10b}$$

where Z_1 and Z_2 are the compressibility factors and R is the gas constant.

The description of the saturation curve ($P = f(T_s)$) for various fluids may be estimated with sufficient accuracy by the following equations:

Nitrogen

$$\text{Log}_{10} P(\text{atm}) = -312.2/T_s + 4.002. \tag{2.11}$$

Oxygen

$$\text{Log}_{10} P = -374.5/T_s + 4.129. \tag{2.12}$$

Helium

$$\text{Log}_{10} P = 0.97864 - 2.7708/T_s + 2.5\text{Log}_{10}T_s. \tag{2.13}$$

For a turbulent single-phase liquid flowing in a channel of diameter D the wall friction force (F) is given by

$$F = \lambda \rho_2^0 v^2/2D, \tag{2.14}$$

where λ is the dimensionless friction factor. In the case of turbulent two-phase flow the wall friction force is higher and can be found by applying the Lockhart–Martinelli correction (Lockhart & Martinelli 1949). This correction was developed for water and has been shown to be valid for two-phase nitrogen and oxygen. Using this correction the wall friction force (F_w) is

$$F_w = M\,F, \tag{2.15}$$

where

$$M = 1 + 20/M_1 + 1/M_1^2, \text{ and} \tag{2.16}$$

$$M_1^2 = (\mu_2/\mu_1)^{0.25}((1 - x)/x)^{1.75}(\rho_1^0/\rho_2^0), \tag{2.17}$$

where x = the quality of the two-phase mixture. The two-phase wall friction is dependent on the physical properties of the phases, the channel geometry, and the mixture quality.

In the case of liquid helium the wall friction force is not described by the Lockhart–Martinelli correction (Boom, El-Wakil, & McIntosh, 1978; Nakagawa et al. 1984). Empirical results from various experiments (Gun & Filippov 1984; Ladohin & Gorbachev 1990; Vishniev, Migalinskaya, & Lebedeva 1982) have been examined by one of the authors (Filina). The parameters of these experiments are listed in Table 2.1. Filina has developed a factor (y) that may be used to estimate the wall friction of two-phase helium. The factor (y) is defined by

$$y = \frac{[(\mu_2/\mu_1)^{0.75}(\rho_1^0/\rho_2^0)]_{He}}{[(\mu_2/\mu_1)^{0.75}(\rho_1^0/\rho_2^0)]_{N_2}}. \tag{2.18}$$

Table 2.1. *Parameters of two-phase helium wall friction experiments*

Parameters	Vishnev et al.	Ladohin et al.	Gun et al.
Channel diameter (mm)	5.2	5.4	4.6
Channel length (m)	1.2	13	2.0
Pressure (bar)	1.3–1.4	1.4	1.0–1.8
Flowrate (kg/m^2 s)	65–400	30–50	100–400

Now the wall friction force for two-phase helium is given by

$$(F_w)_{He} = ((F_w)_{N_2})/y. \tag{2.19}$$

It is important to keep in mind that the expression for y is really only valid for the range of parameters shown in Table 2.1. The effect of changing channel diameter is particularly significant because different flow regimes may be developed in larger channels under the same flow conditions. The extrapolation of wall friction force from smaller to larger channels is not well understood in two-phase helium flow. The expression for helium wall friction given in Equations (2.18) and (2.19) agrees with the measured values to within 30 percent. The value of y under the conditions studied (Table 2.1) is greater than 1. Thus, the wall friction for two-phase helium is less than for that of other cryogens under these conditions. This is also shown by recent experiments (Huang 1994).

As we are modeling the case of equilibrium droplet flow we can assume that all the wall heat flux (Q_w) goes into converting the liquid droplets into vapor. Thus the rate of phase transition ($J_{1,2}$) is given by $J_{1,2} = Q_w/\zeta$, where ζ is the heat of vaporization.

Keeping in mind that ($\rho_1^0 \alpha_1 v + \rho_2^0 \alpha_2 v$) is constant, Equation (2.8) can be expanded to

$$\alpha_1 v(\partial \rho_1^0/\partial P)dP/dz + \rho_1^0 v \, \partial \alpha_1/\partial z + \rho_1^0 \alpha_1 \partial v/\partial z = Q_w/\zeta, \tag{2.20a}$$

$$-\rho_2^0 v \partial \alpha_2/\partial z + \rho_2^0 \alpha_2 \partial v/\partial z = -Q_w/\zeta, \tag{2.20b}$$

$$dP/dz + (\rho_1^0 \alpha_1 + \rho_2^0 \alpha_2)\partial v/\partial z = -F_w. \tag{2.20c}$$

These form a set of three equations with three unknowns defined by

$$B_1 = \partial v/\partial z; \qquad B_2 = \partial \alpha_1/\partial z; \qquad B_3 = \partial P/\partial z. \tag{2.21}$$

Note that

$$\partial \alpha_2 / \partial z = -\partial \alpha_1 / \partial z.$$

In order to numerically integrate Equation (2.20) we must write the unknowns in more explicit terms. From (2.9a) we can write:

$$\partial \rho_1^0 / \partial P = (T_s - P \langle \partial T_s / \partial P \rangle) / (R(T_s)^2), \tag{2.22}$$

where

$$\partial T_s / \partial P = 1/(P \, P^* \langle \ln P^*/P \rangle^2). \tag{2.23}$$

Define $\partial \rho_1^0 / \partial P = A_1$. Then, from Equation (2.20):

$$B_1 = \left(\frac{Q}{\zeta} \left(1 - \rho_1^0 / \rho_2^0 \right) + \alpha_1 \, v \, A_1 (F_w + \rho_m \, g) \right) / (\rho_1^0 - \alpha_1 \, v \, A_1 \rho_m); \tag{2.24a}$$

$$B_2 = (1/\rho_2^0 v) \left[\frac{Q}{\zeta} + \rho_2^0 (1 - \alpha_1) B_1 \right]; \tag{2.24b}$$

$$B_3 = -\rho_m \, B_1 - F_w - \rho_m \, g. \tag{2.24c}$$

Now if the fluid conditions at the entrance to the channel are known, we can use Equation (2.24) to perform an Euler integration in z. This results in

$$v_{i+1} = v_i + B_1 \, \Delta z; \qquad (\alpha_1)_{i+1} = (\alpha_1)_i + B_2 \, \Delta z; \qquad P_{i+1} = P_i + B_3 \, \Delta z. \tag{2.25}$$

Of course, the grid spacing (Δz) will have to be chosen fine enough so that the change in properties within each interval is not too large. Also B_1, B_2, and B_3 are updated according to Equation (2.24) as the numerical integration proceeds. Since the system under study is operating under saturation conditions, a knowledge of the pressure distribution in the channel gives the channel temperature distribution via Equation (2.13).

Thus, knowing the entry conditions of the flow and the wall heat flux (Q_w) it is possible in the case of equilibrium flows to calculate the main parameters of the flow (quality, velocity, and pressure) as a function of steam-generating channel length. The results of these calculations are shown in Figure 2.5. The results of the model calculations are in good agreement with the experimental results (Kuzmenko 1979). Note that the channel length has been divided into three regions: I = the inlet section of nonequilibrium two-phase flow; II = equilibrium droplet flow (in which Equation (2.20) is valid); and III = the one-phase gas flow.

As can be seen, the solution of equilibrium two-phase flows does not require a detailed knowledge of the flow regime. Reliable results from this analysis are obtained if the velocities and temperatures of the phases are indeed equal to or close to each other; that is, $v_1 = v_2$, $T_1 = T_2$. If there are large differences between these parameters, then answers obtained using the equilibrium model are incorrect and a detailed knowledge of the flow structure is required. However, even in the case of obviously nonequilibrium flows, solutions assuming equilibrium conditions may provide a useful estimate as a first step in the analysis of the problem.

In the specific example of nonequilibrium droplet flow within a superheated vapor, which may occur at high flowrates in the gasifier, the wall heat transfer may be described by analogy to the case of water droplets within a superheated flow. The water/steam studies show that the Nusselt (Nu) number should take the form:

$$Nu = A \ Re^m Pr^n f(x, \rho_1^0/\rho_2^0, T/T_w). \tag{2.26}$$

Comparing the results of various researchers (Delhaye et al. 1981; Stirikovich et al. 1982) the formula of Miropolsky is found to be the best. This correlation gives the Nusselt number as

$$Nu = (0.608/x^{1.4})(Pr_w \ Re_1)^{0.8}(x + (\rho_1^0/\rho_2^0)(1 - x)^{0.5}y, \tag{2.27}$$

where

$$Re_1 = \rho_1^0 v \ D/\mu_1; \quad Pr_w = \mu_1 \ C_{p1}/k_1;$$
$$y = 1 - 0.1(\rho_2^0/\rho_1^0 - 1)^{0.4}(1 - x)^{0.4}.$$

The convective heat-transfer coefficient (h) is then related to the Nusselt number by $Nu = hz/k$. The wall heat flux is then defined by

$$Q_w = h(T_w - T). \tag{2.28}$$

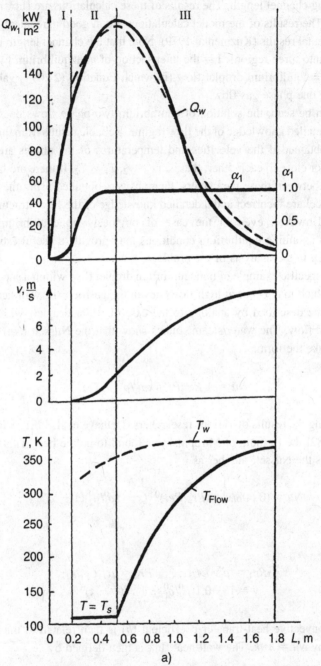

Figure 2.5. (a) Variation of main parameters along gasifier channel length:
- - - - - - - = experimental data (Kuzmenko 1979); ———— = calculation.

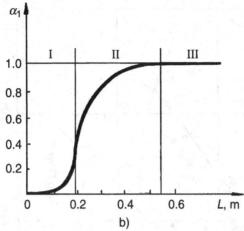

b)

Figure 2.5. (b) Void fraction as a function of channel length as calculated by model: I = nonequilibrium two-phase flow; II = equilibrium droplet flow; III = one-phase gas flow.

2.4 Two-phase helium in magnet-stabilization channels

The principal advantage of using two-phase helium for stabilizing superconducting magnets is that the heat deposited in the helium goes toward changing the phase of the fluid. Thus, the two-phase flow is an essentially isothermal heat sink that maintains the magnets at close to the saturation temperature of the helium. This advantage is so significant that several very large superconducting magnet systems associated with particle accelerators use this cooling technique. These machines include the currently operating Tevatron (Rode, Brindza, Richied, & Stoy 1980) and HERA (Barton et al. 1986) accelerators as well as the proposed Large Hadron Collider (LHC) (Bézaguet et al. 1993; Lebrun 1994). The cooling system of the LHC will include a forced flow two-phase superfluid helium loop.

As an example, consider the two-phase flow found in the superconducting magnets of the UNK accelerator under construction in Russia. In these magnets, two-phase helium at a pressure of 0.13–0.14 MPa flows through pipes (inner diameter = 57 mm) at rates of 200–300 kg/h for distances of 600–700 m. In order to describe the behavior of the two-phase flow in these magnets it is first necessary to determine what flow regimes will be found in the helium under these conditions.

Experimental studies with air/water and steam/water flows have resulted in the Baker diagram (Baker 1954). In this diagram, the various flow regimes (e. g., stratified, wave, slug) are mapped as a function of parameters that take

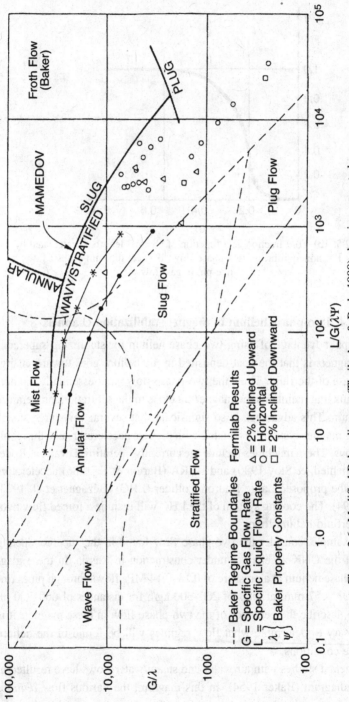

Figure 2.6. Modified Baker diagram (Theilacker & Rode 1988) showing expected operating regions of the UNK magnet cooling channels (D = 57 mm). * = G = 300 kg/h; ● = G = 150 kg/h.

into account the properties of the phases. Other studies (Mamedov 1984; Theilacker & Rode 1988) have shown that in the case of two-phase helium flow, the flow regimes shown on the Baker diagram are effectively shifted to the right of the diagram. Figure 2.6 shows the Baker diagram along with the corrections for two-phase flow found by Mamedov. Also plotted on the figure are the operating ranges for the UNK two-phase system as a function of flow rate and quality ($0.1 \leq x \leq 0.95$). The results indicate that in most cases the wavy stratified flow will be seen in the UNK system.

Because of the different densities of the liquid and vapor phases in the stratified flow, the vapor moving in the upper part of the pipe "slips" relative to the liquid. Figure 2.7 shows the velocity profiles of the phases along the pipe cross section as a function of liquid level. This description is based on physical considerations: the "slippage" of the phases, the retardation "sticking" of the phases at the pipe wall, and the equality of the shear stresses at the liquid–vapor interface (phase boundary). The slippage of the vapor results in the additional turbulization of the flow at the phase boundary causing wave formation and the transport of liquid droplets into the vapor space (droplet carryover). These effects tend to reduce the slippage and equalize the phase temperatures. However, as the liquid level decreases, the surface area of the liquid–vapor interface shrinks and the surface effects of wave formation and droplet carryover become less significant. This results in a superheating of the vapor relative to the saturation temperature of the liquid. Figure 2.7 illustrates the effect of liquid level on the temperature variation along the cross section of the two-phase channel.

The superheating of the vapor in stratified flow becomes more significant as the tube diameter increases. In smaller diameter tubes, thermal conduction along the circumference of the tube wall as well as droplet carryover tend to reduce the slippage between the phases and work to equilibrate the temperature distribution along the tube cross section. At the large-channel diameters envisioned for the UNK machine, analyses (Budrik 1990; Gun, Filippov, & Zinchenko 1985) suggest that the slippage (v_1/v_2) ranges between 2 and 3 and that the vapor may be superheated on the order of hundreds of milliKelvin. This amount of superheat can impact the stabilization of the superconducting magnets. To better understand the performance of the two-phase channels planned for the UNK machine, experimental studies were conducted.

2.4.1 Experimental studies

Previous two-phase horizontal helium flow experiments (Katheder & Süßer 1988) have been carried out with tubes of 10 mm diameter. These researchers

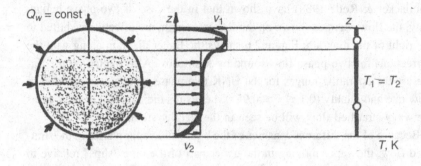

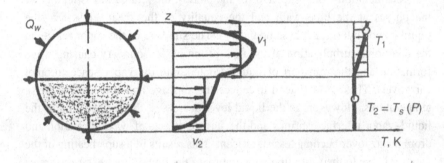

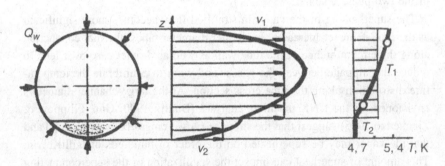

Figure 2.7. Expected two-phase velocity and temperature distributions in a magnet cooling channel as a function of liquid level.

reported slip ratios (v_1/v_2) of 1.5–2.2. They also report that wavy flow was the typical flow regime observed during the experiment.

Since the superheating of the vapor relative to the equilibrium liquid temperature is dependent on the pipe diameter, it was decided to carry out additional experiments using full-size (56 mm ID) UNK two-phase pipes (Filina 1992). The experiments were conducted at the Cryogenmash Company test facility located at Vidnoye, just south of Moscow. The experiment consisted of three 100 m lengths of 56 mm ID pipe connected in series. Each length had a different type of thermal insulation, and thus a different wall heat flux. The helium flow was provided by a refrigerator of the КГУ type, which was equipped with an automated control and data acquisition system. Figure 2.8 is a schematic of the experiment while Figures 2.9–2.11 are photographs of the test facility.

Temperature measurements were made using ТПК–720 resistance thermometers which have a sensitivity of 1 mk at 4.2 K and a time response of no more than 80 ms. These thermometers were mounted vertically down the cross section of the pipeline to provide an accurate measurement of the temperature of both the liquid and vapor phases.

The helium flow provided by the КГУ refrigerator was varied between 0 and 60 kg/h. The flowrate was measured at the refrigeration plant. Once steady-state conditions were reached, the helium temperature, pressure, and flowrate were continuously measured by the data acquisition system. The heat leak into each of the lengths of pipe was determined by closing valves at each end of the pipe length and measuring the subsequent pressure rise as the helium trapped between the valves boiled.

Figure 2.12a shows the position of two thermometers located to measure the vapor and liquid temperature as well as the calculated liquid levels corresponding to flowrates of 60 kg/h and 31 kg/h. Results of temperature measurements are shown in Figure 2.12b. The temperature distribution along the pipe cross section varies as a function of liquid level, which is in turn controlled by the helium flowrate. If the liquid in the pipe occupies the majority of the cross section the amount of vapor superheat is close to or equal to zero. Thus, the vapor temperature is equal to the liquid saturation temperature, which is in equilibrium with the helium pressure. Even at liquid levels of slightly less than half the pipe cross section, the vapor superheat is minimal because the surface area of the vapor liquid interface is comparable to the channel diameter and surface effects such as wave formation and droplet carryover serve to equilibrate the temperatures. However, as the liquid level drops significantly below half the channel diameter, the vapor superheat increases as shown by the data in Figure 2.12b. Once all the liquid in the line evaporates, the temperature in the vapor equilibrates as shown by line 6 in Figure 2.12b. In

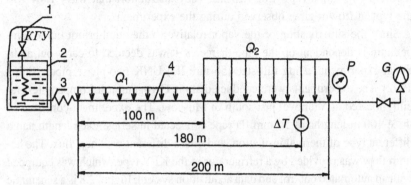

Figure 2.8. Schematic of experimental facility designed to measure the performance of two-phase flow in full-size UNK cooling channels: (1) helium flow from refrigerator, (2) helium bath, (3) inlet line, (4) test section. T = location of temperature sensor; P = location of pressure sensor; G = location of flowmeter; Q_1, Q_2 = constant wall heat flux.

Figure 2.9. Test sections of UNK two-phase flow experimental facility: $L = 100$ m; $D = 52$ mm. Vidnoye, Russia.

pipelines with the diameter of those proposed for the UNK magnets, the superheat can reach its limiting value, defined as the difference between the liquid saturation temperature and the increase of the vapor temperature (T_1) due to the constant wall heat flux ($T_1(z) = Q_w z/(\rho_1 C_p v_1)$). The measured super-

Figure 2.10. Inlet line of UNK two-phase flow experimental facility. Vidnoye, Russia.

Figure 2.11. Close-up of ТПК–720 transducer installation. UNK two-phase flow experimental facility. Vidnoye, Russia.

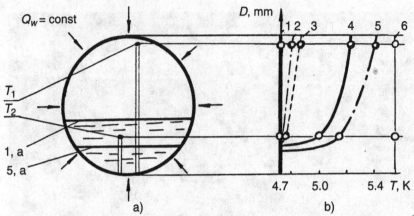

Figure 2.12. Schematic of thermometer location in test section and temperature variation as a flowrate. (a) 1a = calculated liquid helium level at $G = 60$ kg/h; 5a = calculated liquid helium level at $G = 31$ kg/h; T_1, T_2 = thermometer positions. (b) (1) $G = 60$ kg/h; (2) $G = 49$ kg/h; (3) $G = 37$ kg/h; (4) $G = 34$ kg/h, (5) $G = 31$ kg/h; (6) $G = 23$ kg/h.

heat in the experiment was seen to be as much as 0.6–0.8 K above the liquid helium temperature. This temperature rise can result in the quenching (loss of superconductivity) in the UNK magnets. Thus, during accelerator operations care must be taken that the helium flowrates are high enough to avoid low liquid levels in the two-phase pipes.

2.4.2 *Mathematical modeling*

In order to generalize the experimental results, as well as to gain a better understanding of the physics of the problem, a model of the problem was developed. Although the degree of observed vapor superheat is significant for the performance of the superconducting magnets, it is in fact small ($\Delta T/T_{in} = 0.14$) compared to the liquid inlet temperature. Thus, for modeling purposes it is reasonable to begin by assuming that the liquid and vapor temperatures are equal and that all the wall heat, Q_w, is absorbed by the phase transition. The effect of the vapor superheat will be added to the model at a later stage by the use of empirical results. This assumption allows the use of the double-velocity model. In this treatment the phase temperatures are both taken to be equal to the saturation temperature ($T_1 = T_2 = T_s(P)$) and the phase velocities are assumed to be different ($v_1 \neq v_2$). This model is thus somewhat more complicated than the model of equilibrium flow used in Section 2.3.1. Applying the double-velocity model assumptions to Equation (2.5) yields:

Conservation of mass

$$\partial(\rho_1^0 \alpha_1 v_1)/\partial z = Q/\zeta; \tag{2.29a}$$

$$\partial(\rho_2^0 \alpha_2 v_2)/\partial z = -Q/\zeta. \tag{2.29b}$$

Conservation of momentum

$$\partial(\rho_1^0 \alpha_1 v_1^2)/\partial z = -\alpha_1 \partial P/\partial z + F_{1,2} + (Q/\zeta)v_{1,2} - F_{w\,1}; \tag{2.29c}$$

$$\partial(\rho_2^0 \alpha_2 v_2^2)/\partial z = -\alpha_2 \partial P/\partial z - F_{1,2} + (Q/\zeta)v_{1,2} - F_{w\,2}. \tag{2.29d}$$

The conservation-of-mass equations allow the conservation of momentum equations to be rewritten in the form:

$$v_i \, \partial(\rho_i^0 \alpha_i v_i)/\partial z + \rho_i^0 \alpha_i v_i \partial v_i/\partial z = (-1)^{i+1} v_i Q/\zeta + \rho_i^0 \alpha_i v_i \partial v_i \backslash \partial z = f(z). \tag{2.30}$$

Making this substitution, we can expand Equation (2.29) to yield:

$$\alpha_1 v_1 (\partial \rho_1^0/\partial T)dT/dz + \rho_1^0 v_1 \partial \alpha_1/\partial z + \rho_1^0 \alpha_1 \partial v_1/\partial z = Q/\zeta; \tag{2.31a}$$

$$\alpha_2 v_2 (\partial \rho_2^0/\partial T)dT/dz + \rho_2^0 v_2 \partial \alpha_2/\partial z + \rho_2^0 \alpha_2 \partial v_2/\partial z = -Q/\zeta; \tag{2.31b}$$

$$v_1 Q/\zeta + \rho_1^0 v_1 \alpha_1 \partial v_1/\partial z = -\alpha_1 (\partial P/\partial T)dT/dz - F_{1,2} + Qv_{1,2}/\zeta - F_{w\,1}; \tag{2.31c}$$

$$-v_2 Q/\zeta + \rho_2^0 v_2 \alpha_2 \partial v_2/\partial z = -\alpha_2 (\partial P/\partial T)dT/dz + F_{1,2} - Qv_{1,2}/\zeta - F_{w\,2}. \tag{2.31d}$$

Since this is a saturated mixture the relationship between pressure and temperature is given by Equation (2.13) and the densities of the phases may be given by

$$\rho_1^0 = (P(T_s)/Z_1(T_s)R\ T_s); \qquad Z_1 = 0.1456T_s; \tag{2.32}$$

$$\rho_2^0 = Z_2 T_s; \qquad Z_2 = 27.3. \tag{2.33}$$

Thus,

$$\frac{\partial \rho_1^0}{\partial T} = \frac{P' Z_1 T_s - P Z_1' T_s - P Z_1}{R Z_1^2 T_s^2},$$

$$(P' = \partial P/\partial T; \; Z_1' = \partial Z_1/\partial T), \tag{2.34}$$

and

$$\frac{\partial \rho_2^0}{\partial T} = Z_2. \tag{2.35}$$

Defining

$$A_1 = \partial \rho_1^0/\partial T; \qquad A_2 = \rho_1^0 \alpha_1 + \rho_2^0 \alpha_2; \qquad A_3 = \partial P/\partial T,$$

Equation (2.31) can be written as

$$\alpha_1 v_1 A_1 dT/dz + \rho_1^0 v_1 d\alpha_1/dz + \rho_1^0 \alpha_1 dv_1/dz = Q/\zeta; \tag{2.36a}$$

$$\alpha_2 v_2 Z_2 dT/dz - \rho_2^0 v_2 d\alpha_2/dz + \rho_2^0 \alpha_2 dv_2/dz = -Q/\zeta; \tag{2.36b}$$

$$\alpha_1 A_3 dT/dz + \rho_1^0 v_1 \alpha_1 dv_1/dz = -v_1 \, Q/\zeta - F_{1,2} + v_{1,2} \, Q/\zeta - F_{w\,1}; \tag{2.36c}$$

$$\alpha_2 A_3 dT/dz + \rho_2^0 v_2 \alpha_2 dv_2/dz = v_2 \, Q/\zeta + F_{1,2} - v_{1,2} \, Q/\zeta - F_{w\,2}. \tag{2.36d}$$

This system of four equations with four unknowns (dv_1/dz, dv_2/dz, $d\alpha_1/dz$, dt/dz) may be solved analytically to yield:

$$dv_1/dz = B_3 - B_4 dT/dz, \tag{2.37a}$$

$$dv_2/dz = B_5 - B_6 dT/dz, \tag{2.37b}$$

$$d\alpha_1/dz = B_2 + B_1 dT/dz, \tag{2.37c}$$

$$\frac{dT}{dz} = \frac{(Q/\zeta) - \rho_1^0 v_1 B_2 - \rho_1^0 \alpha_1 B_3}{\alpha_1 v_1 A_1 + \rho_1^0 v_1 B_1 - \rho_1^0 \alpha_1 B_4}, \tag{2.37d}$$

where

$$B_1 = \frac{1}{\rho_2^0 v_2} \left(\alpha_2 v_2 Z_2 - \frac{\alpha_2 A_3}{v_2} \right), \tag{2.38a}$$

$$B_2 = \frac{1}{\rho_2^0 v_2^2} \left(v_2 \frac{Q}{\zeta} + F_{1,2} - v_{1,2} \frac{Q}{\zeta} - F_{w\,2} \right) + \frac{1}{\rho_2^0 v_2} \frac{Q}{\zeta}, \tag{2.38b}$$

$$B_3 = \frac{1}{\rho_1^0 v_1 \alpha_1} \left[\frac{Q}{\zeta} (v_{1,2} - v_1) - (F_{1,2} + F_{w\,1}) \right], \tag{2.38c}$$

$$B_4 = \frac{\alpha_1 A_3}{\rho_1^0 v_1 \alpha_1}, \tag{2.38d}$$

$$B_5 = \frac{1}{\rho_2^0 v_2 \alpha_2} \left[\frac{Q}{\zeta} (v_2 - v_{1,2}) + F_{1,2} - F_{w2} \right], \tag{2.38e}$$

$$B_6 = \frac{\alpha_2 A_3}{\rho_2^0 v_2 \alpha_2}. \tag{2.38f}$$

Now we still must determine the friction between each phase and the wall (F_{wi}) as well as the friction forces at the liquid–vapor interface.

Since the flow is stratified, the specific friction force (N/m^3) between each phase and the wall may be treated as a separate single-phase problem. Thus, the friction force is found by substituting the velocity of each phase into the single-phase Blasius correlation for the friction factor (λ). The resulting wall friction is given by

$$F_{wi} = (\lambda_i \rho_i^0 v_i^2 L_i)/2S, \tag{2.39}$$

where

$$\lambda_i = (0.3164)/Re_i^{0.25} \quad \text{(Blasius correlation)}, \tag{2.40}$$

$$Re_i = \rho_i^0 v_i D_i / \mu_i, \tag{2.41}$$

$S =$ the total cross-sectional flow area $= (\pi D^2/4)$,
$L_i =$ the perimeter covered by the ith phase,
$D_i = L_i/\pi; \; D_1 + D_2 = D$.

The liquid–vapor interface friction force ($F_{1,2}$) resulting from the velocity mismatch at the interface may be treated in a similar manner. Here the relevant velocity is the longitudinal velocity of the mass undergoing evaporation. This velocity ($v_{1,2}$) is greater than the liquid velocity (v_2) but less than the vapor velocity (v_1). For engineering applications this velocity can be taken as the average of the liquid and vapor velocities. Thus, $v_{1,2} = (v_1 + v_2)/2$. The friction force is given by

$$F_{12} = (\lambda_{12}\rho_i^0 v_{1,2}^2 L_{1,2})/2S, \tag{2.42}$$

where $\lambda_{12} = (0.3164)/Re_{1,2}^{0.25}$ and $Re_{1,2} = \rho_1^0 v_{1,2} D_1/\mu_1$.

$L_{1,2}$ is the chord length of the interface. Note that all the property values used in the calculation are those of the gas phase.

The use of the traditional physical concepts of fluid friction to determine F_{wi} and $F_{1,2}$ make it possible to complete the model of the stratified flow without invoking any empirical relationships for the slippage between the phases.

When analyzing a stabilization channel, the following inlet conditions are generally known: temperature (T); pressure (P); wall heat flux (Q_w); and void fraction (α_1). From the value of α_1, the inlet values of α_2, L_1, L_2, D_1, and D_2 may be found. The inlet physical properties (density, viscosity, etc.) may, of course, be calculated from the inlet pressure and temperature. The inlet velocities v_1 and v_2 may be calculated by iteration of Equation (2.37) with the requirements that $dv_i/dz > 0$ at the inlet and that the total mass is conserved.

Equation (2.37) may be combined with the inlet conditions and Euler integration (e.g., $v_1(z_{i+1}) = v_1(z_i) + \Delta z\, dv_1/dz$) to calculate $v_1(z)$, $v_2(z)$, $\alpha_1(z)$, and $T(z)$ along the length of the channel. As always in Euler integration the grid size (Δz) must be made small enough to avoid error.

Recall that one of the assumptions of the model is that no vapor superheat was present. Therefore, $T_1(z) = T_2(z)$. But, as previously described, experiments have shown that the vapor can become superheated. The amount of this superheat cannot be easily determined from first principles and data from experiments must be used. In the experiments described in Section 2.4.1, the average vapor superheat was found to be

$$\Delta \overline{T} = Q_w D_1/(Nu_{1,2}\lambda_{1,2}). \tag{2.43}$$

The Nusselt number ($Nu_{1,2}$) has been found in these experiments to equal

$$Nu_{1,2} = (2.5 \times 10^{-3}/\alpha_1^{0.8})(Re_{1,2}^{0.8})(Pr_1^{0.4}), \qquad (2.44)$$

where $Pr_1 = c_{p1}\mu_1/K_1$ and K_1 is the thermal conductivity of the vapor.

These equations agree with the experimental data to within 35 percent. They are valid for $0.1 - 0.5 \le \alpha_1 \le 0.95$ and $10^4 \le Re < 10^6$. Thus, throughout the cryostabilization channel $T_1(z) = T_2(z) + \overline{\Delta T}$.

The results of the model calculations and the experiments are shown in Figures 2.13 and 2.14. As the flowrate drops, the stratified flow level drops in the pipe cross section and both the vapor superheat and the void fraction increase. Figure 2.13a shows good agreement between the measured vapor superheat and the superheat calculated by Equation (2.43). This figure is divided into two regions: I, single-phase gas flow, and II, two-phase stratified flow. The maximum superheat indicated on the figure is calculated by assuming that all the wall heat flux goes into raising the vapor temperature, with the result that $dT_1(z)/dz = Q_w/(\rho_1 C_p v_1)$. This, of course, occurs at the beginning of the single-phase region. In Figure 2.13b, the variation of void fraction as function of mass flowrate is shown. The void fraction was calculated using the mathematical model (Equation 2.36). Note that the void fraction calculated by the model approaches 1 at the same flowrate for which the experimental data in Figure 2.13a indicate that single-phase flow is reached. This consistency between the data and the model gives us confidence in the correctness of the model. The slip (v_1/v_2) calculated by the model is compared to the results of empirical formulas of various authors (Budrik et al. 1990; Filippov, Mamedov, & Selyunin 1988; Gun, Filippov, & Zinchenko 1985) in Figure 2.13c. There is general agreement between these results. Figure 2.14 shows the change in velocity, temperature, and void fraction along the length of the experimental channel at a flowrate of 31 kg/hr as calculated by the model. As the void fraction increases, the slippage between the vapor and liquid phases increases, as does the vapor superheat. The vapor velocity approaches the single-phase gas velocity ($v_g = G/\rho_1 S$) as the void fraction approaches 1. Notice also, in Figure 2.15b, that the calculated temperatures T_1 and T_2 agree with the experimentally measured values at $L = 180$ m.

The void fraction is the most significant parameter in this problem. The vapor superheat is dependent on the void fraction and experimental results have shown that if the void fraction is greater than approximately 0.7 to 0.8 the superheating of the vapor becomes excessive. Figure 2.15a shows the increase

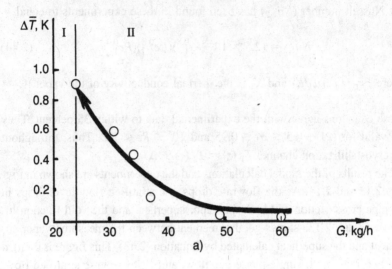

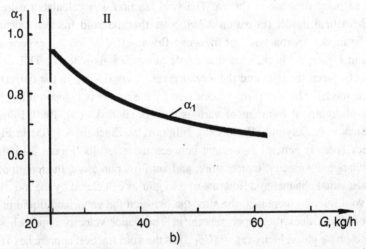

Figure 2.13. Effect of helium flowrate on the fluid parameters. (a) Average superheat of vapor in the test section: I = single-phase gas flow; II = separated two-phase flow; O = data; ——— = calculation according to Equation (2.43); → = maximum vapor superheat in test section. (b) Calculated void fraction in test section.

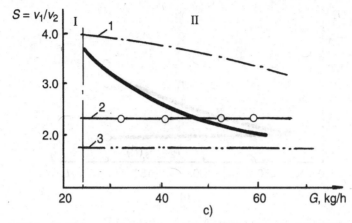

Figure 2.13. (c) Slippage in test section: (1) calculation from Budrik (1990); (2) calculation from Filippov, Mamedov, & Selyunin (1988); (3) calculation from Gun, Filippov, & Zinchenko (1985); ———— = calculation by author (N.F.).

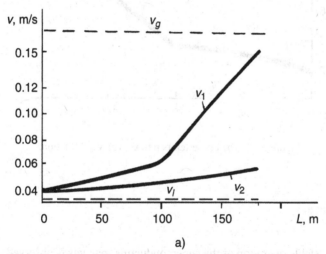

Figure 2.14. Variation of flow parameters as a function of test section length at $G = 31$ kg/h. $\bigcirc$ = data; ———— = calculation according to Equation (2.36). (a) Phase velocities: v_1 = vapor velocity; v_2 = liquid velocity; v_l, v_g = single-phase limiting velocities. (continued on next page)

of the superheat in stratified flow as a function of the void fraction. The limiting case at $\alpha_1 = 1$ is determined by assuming that all the heat goes into raising the single-phase vapor temperature. The results of this work are consistent with the operational experience of the Tevatron accelerator. In the

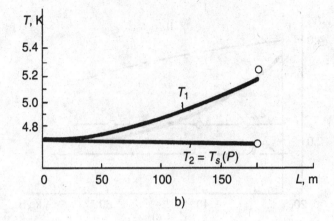

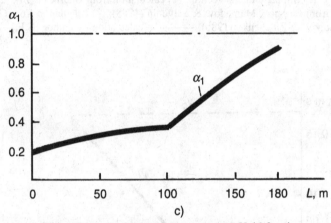

Figure 2.14. (b) Phase temperatures. (c) Void fraction.

Tevatron, stable operation of the superconducting magnets is observed as long as the liquid level in the two-phase cooling channels is not less than 20 percent of the cross section (corresponding to $\alpha_1 = 0.8$). If the void fraction rises above 0.8, the increased vapor superheat affects the magnet performance.

The slippage (v_1/v_2) between the phases is also dependent upon the void fraction. Figure 2.15b compares the slippage as a function of void fraction calculated by Equation 2.40 with calculations from empirical relationships (Budrik et al. 1990; Filippov et al. 1988; Gun et al. 1985). The fact that some of the empirical formulas predict that the slippage is independent of the void fraction

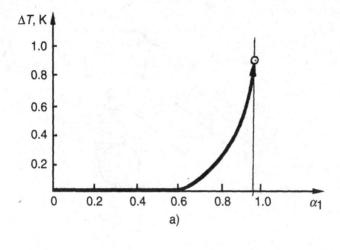

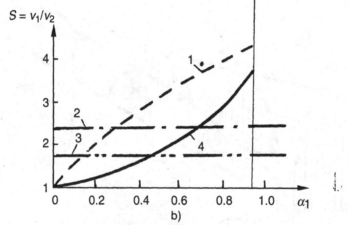

Figure 2.15. (a) Calculated vapor superheat as a function of void fraction in separated two-phase flow: $\bigcirc$ = superheat in the single-phase gas flow. (b) Slippage in separated flow as a function of void fraction: (1) calculation from Budrik (1990); (2) calculation from Filippov et al. (1988); (3) calculation from Gun et al. (1985); (4) calculation by author (N.F.) using two-velocity model.

is difficult to understand. Additional experiments could be carried out to investigate this, but it is really the degree of vapor superheat, not the slippage, that is most important to the proper operation of the superconducting magnets.

A qualitative representation of the change in phase temperatures and velocities caused by changes in the void fraction is shown in Figure 2.16. Notice the discontinuity in the velocity profile at the phase boundary.

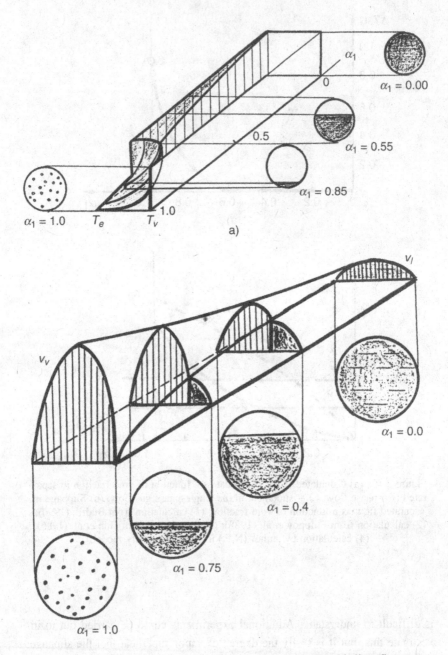

Figure 2.16. (a) Variation of temperature in two-phase separated helium flow as a function of void fraction. (b) Variation of phase velocities in two-phase separated helium flow as a function of void fraction.

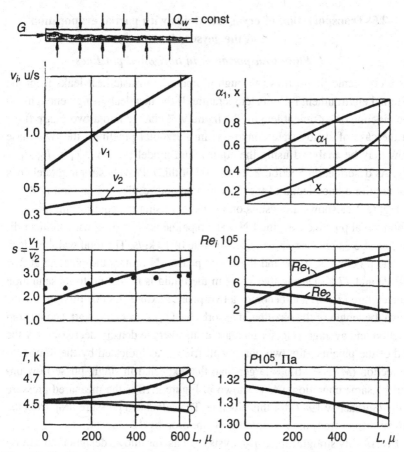

Figure 2.17. Variation of the main parameters in the two-phase line of the UNK superconducting magnets. $G = 300$ kg/hr; $Q = 10$ W/m; $L = 650$ m; $D = 57$ mm; ——— = calculation from author's (N.F.) model. $\bigcirc$ = empirical relation (Equation (2.43)). $\bullet$ = calculation according to Budrik (1990).

The final result of the analysis of the UNK two-phase system is that for the nominal flowrates planned, the vapor superheat at the end of the 650 m magnet chain is 0.21 K. This amount of superheat is allowable and does not cause quenches in the superconducting magnets. Figure 2.17 summarizes the variation of parameters along the length of the UNK magnet chain as calculated by the double-velocity model.

2.5 Transportation of cryogenic fluids with partial evaporation and the geyser effect

2.5.1 Fluid transportation in horizontal pipelines

As a cryogenic liquid flows through a pipeline intrinsic heat leaks from the external environment raise its temperature. If the heat leak is high enough and the pipeline of sufficient length, the liquid will change into a two-phase flow. In the case of steady-state conditions in horizontal pipelines, the two-phase flow may be analyzed using the equilibrium model ($v_1 = v_2$, $T_1 = T_2$, $P_1 = P_2$). As described in Section 2.3.1 the use of this relatively simple model does not require a knowledge of the flow regime and results in Equation (2.8).

Figure 2.18 shows the results of an analysis using the equilibrium model of a horizontal pipeline carrying LN_2. The pipeline is 15 m long with an inner diameter of 2.34 cm carrying a mass flowrate of 10 kg/s. The total wall heat leak for the pipeline is 30 W and the single-phase LN_2 enters the pipeline at 2×10^5 Pa and 77 K. During the first 2 m the liquid is heated up to its saturation temperature ($T_s(P)$) and becomes a two-phase mixture. As this mixture travels down the length of the pipeline, it absorbs heat (Q_w) from the wall, and its void fraction and average velocity increase as its average density decreases. By the end of the pipeline, the two-phase wall friction as indicated by the Martinelli correction (M) is 30 times larger than that expected in single-phase flow under the same conditions. Also shown in Figure 2.18 is the measured pressure drop (Gorbachev 1987) for this pipeline. The calculated pressure drop from the model differs from the measured pressure drop by 12 percent.

Reasonably straightforward analysis such as this allows designers to choose the pipeline geometry required to produce the desired pressure drop, flowrate, and fluid outlet conditions. Note that the equilibrium model neglects both the slippage between the phases and any superheating of the vapor.

2.5.2 The geyser effect

The behavior of fluid transportation systems containing inclined or vertical pipelines and elevated storage tanks is complicated by the geyser effect. This phenomenon, which typically occurs at very low flowrates or with stagnant fluids, is characterized by a sudden boiling of all the fluid in the inclined pipeline and the ejection of the resulting two-phase flow as a geyser into the upper storage tank. Once the inclined pipeline has emptied, fluid from the upper tank quickly refills the pipeline, which can cause damage by a water hammer effect. This process of geysering and refilling can then repeat. In these systems, the principal heat leak into the inclined pipeline comes from the shut-

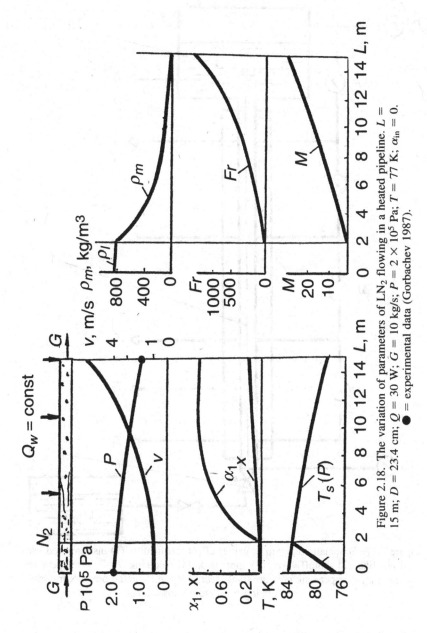

Figure 2.18. The variation of parameters of LN$_2$ flowing in a heated pipeline. $L = 15$ m; $D = 23.4$ cm; $Q = 30$ W; $G = 10$ kg/s; $P = 2 \times 10^5$ Pa; $T = 77$ K; $\alpha_{in} = 0$. ● = experimental data (Gorbachev 1987).

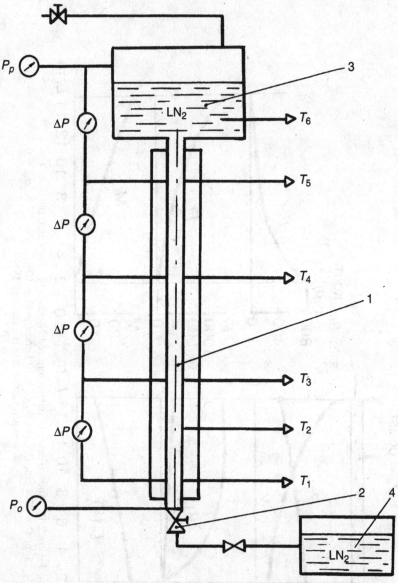

Figure 2.19. Schematic diagram of geyser effect test facility: (1) instrumented verti-cal pipeline, (2) shutoff valve, (3) upper tank, (4) feed tank. T_1–T_6 = temperature sensors; P_p = upper tank pressure transducer; ΔP = differential pressure transduc-ers; P_0 = lower pressure transducers.

off valve at the bottom of the pipe rather than through the pipeline wall. The geyser effect has been observed in systems associated with the filling of cryogenic storage tanks in spacecraft (in both the United States and Russia) as well as in liquefied natural gas storage systems (Johnson 1983).

Experiments (Filina 1983) on the geyser effect using LN_2 have been performed using industrial-scale components. Figure 2.19 is a schematic of the experimental facility. Table 2.2 lists the parameters of the components studied as well as some of the experimental results. In the table and figure P_p is the pressure in the upper tank, Q_o is the heat input at the shutoff valve (item 2 in Figure 2.19), N is number of observed geysers during one run of the experiment, and τ is the average time observed between geysers. A double-walled vacuum insulated glass pipeline with an inner diameter of 0.024 m and a length of 2 m was also used in some of the experiments to permit direct visual observation of the geysering process.

As shown in Table 2.2, it was observed that for each set of pipeline geometries and initial conditions there was a reproducible time τ between the geysers. This suggests that the geyser effect can be understood in terms of these initial conditions and geometries. The number of geysers observed is driven by the initial level in the upper storage tank. The geysers are seen to continue until all the nitrogen in the upper tank and pipeline is converted to vapor.

Figures 2.20 and 2.21 show the observed temperature distribution as a function of time and position along the vertical pipeline for two different end heat inputs. As heat leaks into the pipeline from the shutoff valve at the bottom the temperature of the LN_2 is raised to its saturation temperature (as determined by the local pressure) and nucleate boiling begins. This heated two-phase layer grows in height until it reaches a critical height (H_{cr}) at which point vigorous boiling is suddenly seen in the entire pipeline and the geyser is ejected. This process is illustrated by the data in Figure 2.22.

The explanation behind the sudden explosive boiling and the resulting geyser is related to the flow regime of the two-phase flow in the inclined pipe. Initially, when the nucleate boiling begins, the vapor bubbles rise to the top of the pipe and condense upon entering the cooler upper layers. As the heated two-phase layer grows in height an increasing number of bubbles are formed and start to influence each other and coalesce. The critical height (H_{cr}) is reached when the void fraction of the vapor (α_1) reaches 0.3–0.4. Under these conditions the bubbles merge together to form a slug flow. This vapor slug interacts with the pipe wall and slows down, accumulating more bubbles from below. The growing vapor slug acts to reduce the hydrostatic pressure on the heated lower layers, which in turn results in more boiling. This process then feeds back on itself causing an avalanche of boiling in the entire pipe and the geyser is formed. After the empty tube is refilled from the upper storage tank,

Table 2.2. Test parameters and resulting data for geyser experiments

H (m)	7.5			10.5			12.2					
D (m)	0.1			0.05			0.1					
P_p (MPa)	0.1			0.1			0.1			0.18	0.24	0.28
Q_o (W)	65	100	60	180	200	260	70	140	240		240	
N	11	11	18	22	10	23	10	11	10	9	11	10
τ (s)	703	465	459	145	136	109	1506	731	440	227	134	107

the process will repeat itself if the end heat leak and other initial conditions remain the same. Figure 2.23 illustrates the geyser effect. The differential pressure traces shown in Figure 2.24 show the development of the geyser and the pressure spike caused by the sudden refilling of the pipe from the upper tank.

The geyser effect is a complicated dynamic process. However, for most engineering applications it is sufficient to answer two questions: What is the critical height for the geyser effect to take place under a given set of conditions? And how much time (τ) is there after shutting off the flow in the pipeline before geysering begins? The answer to these questions may be found using a fairly simple model.

Start by assuming that the fraction of end heat leak (Q_s) that goes into the liquid phase in given by

$$Q_s = Q_o(1 - \alpha_1). \tag{2.45}$$

The equilibrium boiling is described by

$$Q_s d\tau = C_{p2} \, d(m_2 \Delta T) + \zeta dm_1, \tag{2.46}$$

where

C_{p2} = the specific heat of the saturated liquid,
m_2 and m_1 = the masses of the saturated liquid and vapor respectively,
$\Delta T = T_s - T_o$,
ζ = the latent heat of the liquid.

The masses (m_1, m_2) are straightforwardly given by

$$m_2 = \rho_2^0(1 - \alpha_1)HS; \quad S = D^2/4; \tag{2.47}$$

$$m_1 = \frac{\rho_1^0}{\rho_2^0}\left(\frac{\alpha_1}{1 - \alpha_1}\right)m_2. \tag{2.48}$$

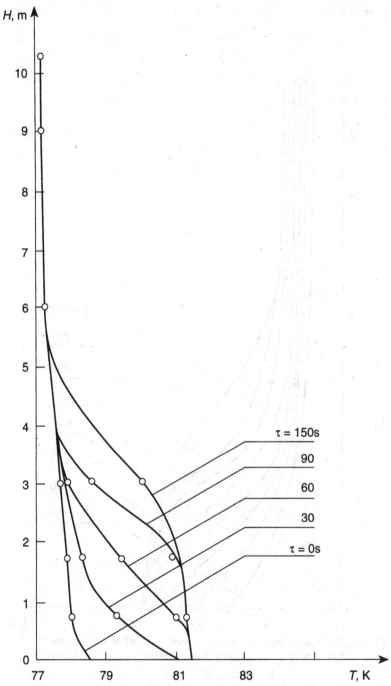

Figure 2.20. Temperature distribution in vertical pipeline with stagnant LN_2; $H = 10.5$ m; $d = 50$ mm; end heat input = 180 W; $\bigcirc$ = experimental data.

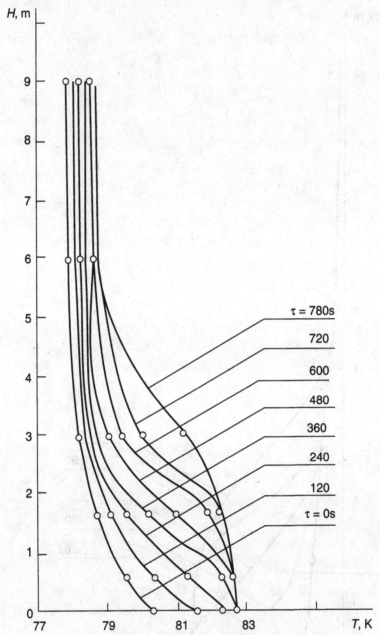

Figure 2.21. Temperature distribution in vertical pipeline with stagnant LN$_2$. $H =$ 12.2 m; $d =$ 100 mm; end heat input $=$ 140 W; $\bigcirc$ = experimental data.

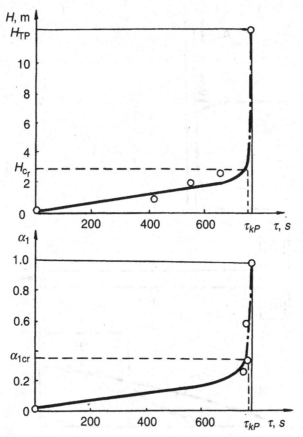

Figure 2.22. The variation of the saturated layer height and its void fraction as a function of time in a vertical column. $H = 12.2$ m; $d = 0.1$ m; $Q = 140$ W; $P = 0.1$ MPa; ——— = calculation, ○ = experimental data.

In this model H and α_1 are functions of time (τ). Since this problem involves fluid under saturation conditions the Clausius–Clapeyron equation may be used to relate the changes in pressure and temperature:

$$dT = \frac{T_s}{\zeta \rho_1^0} dP. \qquad (2.49)$$

The mean pressure within the saturated liquid is given by

$$\overline{P} = P_{\tan k} + \rho_{2g}^0(H_{Pipe} - H_s) + \rho_2^0(1 - \alpha_1)gH_s/2, \qquad (2.50)$$

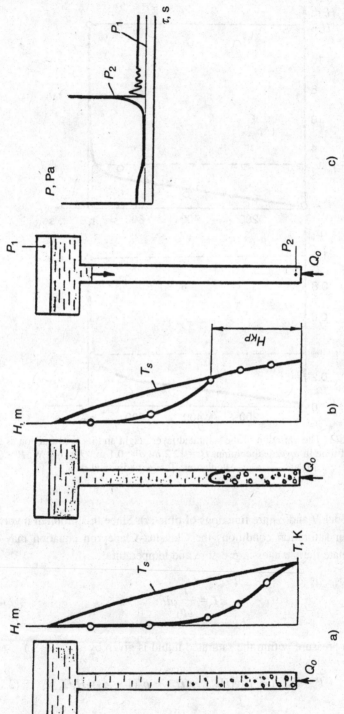

Figure 2.23. The geyser effect: (a) nucleate boiling, (b) slug flow, (c) water hammer during refilling of pipeline from upper tank.

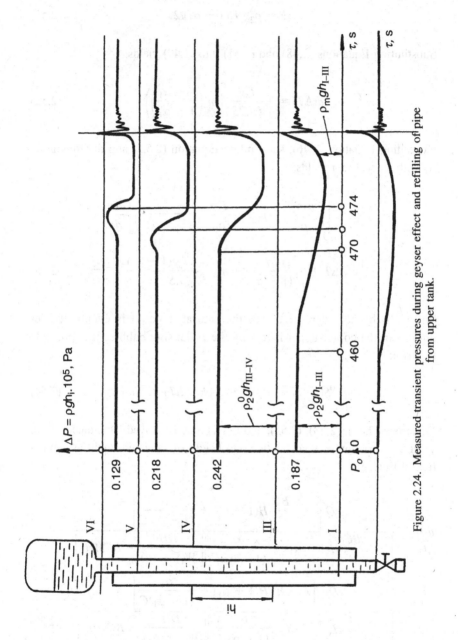

Figure 2.24. Measured transient pressures during geyser effect and refilling of pipe from upper tank.

where H_s is the height of the saturated region and H_{pipe} is the length of the entire vertical pipe. Equation (2.50) yields:

$$\Delta P = \rho_2^0 g \, H(1 + \alpha_1)/2. \tag{2.51}$$

Substituting Equations (2.48) and (2.51) into (2.49) yields:

$$d(\Delta T) = -\frac{T_s g}{\zeta \rho_1^0 2 S} d\left(m_2 \left(\frac{1 + \alpha_1}{1 - \alpha_1} \right) \right). \tag{2.52}$$

Expanding Equation (2.46), substituting Equation (2.52), and differentiating α_1 with respect to τ yields:

$$Q_s = \left[\left(\frac{m_2 \rho_1^0 \zeta}{\rho_2^0} - m_2^2 C_{p2} \frac{T_s g}{\zeta \rho_1^0 S} \right) \left(\frac{1}{(1 - \alpha_1)^2} \right) \right] \frac{d\alpha_1}{d\tau}$$

$$+ \left[C_{p2} \Delta T + \frac{\zeta \rho_1^0 \alpha_1}{\rho_2^0 (1 - \alpha_1)} - m_2 C_{p2} \frac{T_s g(1 + \alpha_1)}{\zeta \rho_1^0 S(1 - \alpha_1)} \right] \frac{dm_2}{d\tau}. \tag{2.53}$$

Since all the heat deposited into the saturated liquid is eventually deposited in the cooler upper layers by the collapsing bubbles it is possible to write:

$$\zeta \rho_1^0 S(H_{pipe} - H_s) d\tau = C_{p2} d(m_2 \Delta T) + \zeta dm_1. \tag{2.54}$$

Equations (2.47), (2.48), (2.53), and (2.54) may be solved simultaneously to yield the variation of void fraction and height of saturated liquid as a function of time:

$$\frac{d\alpha_1}{d\tau} = \frac{Q_s \left[A - \frac{BC}{2} H(1 - \alpha_1) + \frac{\alpha_1}{(1 - \alpha_1)C} \right]}{\left[A - \frac{BC}{2} H(1 - \alpha_1) + \frac{\alpha_1}{(1 - \alpha_1)C} \right] \left[\frac{DH}{W(1 - \alpha_1)} - BCMH^2 \right]}$$

$$- \frac{D\alpha_1 \left[A - BCH(1 + \alpha_1) + \frac{\alpha_1}{(1 - \alpha_1)C} \right]}{\left[A - \frac{BC}{2} H(1 - \alpha_1) + \frac{\alpha_1}{(1 - \alpha_1)C} \right] \left[\frac{DH}{W(1 - \alpha_1)} - BCMH^2 \right]}; \tag{2.55}$$

$$\frac{dH}{d\tau} = \frac{D\alpha_1}{M(1 - \alpha_1)\left[A - \frac{BC}{2}H(1 + \alpha_1) + \frac{\alpha_1}{(1 - \alpha_1)}\right]}$$

$$+ H/(1 - \alpha_1)\frac{d\alpha_1}{d\tau}; \qquad (2.56)$$

where

$$A = C_{p1}\Delta T; \qquad B = C_{p1}gT_s; \qquad C = \frac{\rho_2^0}{\rho_1^0\zeta}; \qquad D = SW\rho_1^0\zeta; \qquad M = S\rho_2^0.$$

These equations are plotted in Figure 2.22 along with the experimental data. Note that there is good agreement between the data and the model. To determine the critical height of the saturated liquid along with the corresponding time before the initiation of the geyser consider that just before the geyser occurs $Q_s = 0$ and $dm_2 = 0$. Making these substitutions into Equation (2.53) and solving for H_{cr} yields:

$$H_{cr} = \frac{\zeta^2}{C_{p2}gT_s(1 - (\alpha_1)_{cr})}\left(\frac{\rho_1^0}{\rho_2^0}\right)^2. \qquad (2.57)$$

The time (τ) then is just the length of time required for the saturated liquid to reach H_{cr} at a given heat leak Q_s. Thus:

$$Q_s\tau = C_{p2}\rho_2^0 H_{cr}S(T_s - T_0). \qquad (2.58)$$

Therefore,

$$\tau = \frac{(\zeta\rho_1^0)^2}{\rho_2^0 gT_s(1 - (\alpha_1)_{cr})4Q_0}(T_s - T_0)\pi D^2. \qquad (2.59)$$

Experimentally, α_{1cr} has been found to range between 0.3 and 0.4. The time until the development of a geyser is inversely proportional to the end heat leak (Q_o). Thus, to avoid geysering for a given diameter pipeline, low-heat-leak valves should be used as the shutoff valves to reduce Q_o and the operation of the storage system should be designed so that stagnant fluid does not sit in the vertical pipeline longer than τ. Failure to take these precautions could result in damage to the system being caused by the geyser and water hammer effects.

2.6 Two-phase flow regimes and optimum heat transfer

Two-phase flow regimes may also be broadly characterized as "regular" or "irregular." Consider Figure 2.25, which shows three different flow regimes. If the phases are well dispersed, and the characteristic size of the inclusions (d_o) and the average distance between inclusions (d) are much less than the channel diameter (D), then the flow is considered regular. If these conditions are not met, the flow is termed irregular. Examples of regular flows include bubble and mist flows. Slug, plug, wavy, and stratified flows are irregular flows.

Regular flows have a distinct advantage in transferring heat from a warm surface. The dispersed heat transfer surface provided by the bubbles or droplets suspended in the flow can be significantly larger than the pipeline containing the flow. Thus, all the wall heat typically goes into converting the liquid into vapor rather than increasing the vapor temperature. This results in more efficient heat transfer.

Flow regimes are generally mapped by empirically determined condition charts (Delhaye et al. 1981; Mandane et al. 1974; Stirikovichet et al. 1982) such as the modified Baker diagram seen in Figure 2.6. Using these charts, designers seeking optimal wall heat transfer can adjust the parameters of the two-phase flow and the system geometry in an attempt to create regular flow inside the system. This may be done at the expense of a more complicated design or higher pressure drops. Design trade-offs may have to be made and regular flows may not be possible in all systems.

The design of the gasifier described in Section 2.3 sought to create a dispersed droplet flow (a regular flow) by the use of staggered fins. The presence of these increased the pressure drop, but the resulting regular flow did produce an overall more efficient gasifier. The stratified flow in the magnet cooling channels of Section 2.4 is an irregular flow and results in the superheating of the vapor component. This superheating is kept to a manageable level by ensuring that the operational mass flowrate is large enough so that the liquid level in the channel does not become too small. In effect, the solution is to make the flow appear more regular.

The geyser effect may also be viewed in terms of regular and irregular flows. As long as dispersed nucleate boiling is taking place in the vertical pipe, heat is transferred smoothly from the base of the pipe into the LN_2. However, once the irregular slug flow is formed, the geyser effect and its associated problems occurs.

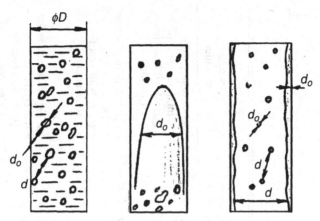

Figure 2.25. Characteristic dimensions of two-phase flow in a channel with diameter D. d_o = average dimension of inclusions; d = average distance between inclusions.

2.7 Modeling of two-phase flows

Table 2.3 lists the principal classes of two-phase flow models and their characteristics. The models range from the fairly simple, such as the equilibrium model that is described in three equations, to the very complex – the three-velocity, two-temperature model that requires the simultaneous solution of nine equations. Selection of the proper model to use for a given physical problem is one of the first steps in modeling two-phase flow. The goal is to pick the simplest model that includes all the relevant physical phenomena.

One very effective way to select the proper model is to construct a simplified sketch of the problem showing the important velocities, temperatures, heat flows, interactions between the phases, and interactions between the phases, and the wall. A series of these sketches for a variety of practical problems is given in Figure 2.26.

Figure 2.26a shows fully developed nucleate boiling in the flow. The bubbles are uniformly dispersed in the liquid, the phases have the same velocity and temperature, and all the external heat goes into the phase transition. This situation can be analyzed using the equilibrium model. This model is also appropriate for the situation described in Figure 2.26b, which shows the vapor–droplet flow found in the cryogenic gasifier described in Section 2.3. The case of stratified two-phase flow in magnet cooling channels, which was addressed in Section 2.4, is sketched in Figure 2.26c. Recall that in this problem the velocities of the phases are different and the temperatures of the phases are taken to be equal. The superheating of the vapor was added later

Table 2.3. Comparison of various two-phase flow models

Description of model	Number of equations	Conservation equations	Description of phase transition	Independent variables	Interface quantities
Equilibrium model $v_1 = v_2$; $T_1 = T_2$;	3	2 Continuity 1 Momentum	$J_{12} = Q_w/\zeta$	α_1; P; v	F_w = Wall friction for mixture. Q_w = Wall heat flux for mixture.
Equilibrium model (with large pressure drops) $\Delta P \sim P_0$; $L \gg D$; Re = $10^4 - 10^7$	4	2 Continuity 1 Momentum 1 Energy	—	α_1; P; v; J_{12}	F_w = Wall friction for mixture. Q_w = Wall heat flux for mixture.
Two-velocity model $v_1 \neq v_2$; $T_1 = T_2 = T_s(P)$;	4	2 Continuity 2 Momentum	$J_{12} = Q_w/\zeta$	α_1; P; v_1; v_2	F_w = Wall friction for mixture. Q_w = Wall heat flux for mixture. F_{12} = Interface friction.

5	Two-temperature model $v_1 = v_2$; $T_1 > T_2 = T_s (P)$;	2 Continuity 1 Momentum 2 Energy	—	α_1; P; v; T_1; J_{21}	F_w = Wall friction for mixture. Q_{wi} = Wall heat flux for ith phase. Q_{ij} = Heat flux between phases.
6	Two-velocity, Two-temperature model $v_1 \neq v_2$; $T_1 > T_2 = T_s (P)$;	2 Continuity 2 Momentum 2 Energy	—	α_1; P; v_1; v_2; T_1; J_{21}	F_w = Wall friction for mixture. Q_{wi} = Wall heat flux for ith phase. Q_{ij} = Heat flux between phases. F_{12} = Interface friction.
9	Three-velocity, Two-temperature model v_1, v_2, v_3; $T_1 > T_2 = T_s(P)$; $T_3 = T_2$	3 Continuity 3 Momentum 3 Energy	—	α_2 (droplets); α_2 (film); P; v_1; v_2; v_3; T_1; J_{21}; J_{31}	F_{w3} = Wall friction for film. Q_{w3} = Wall heat flux for film. Q_{ij} = Heat flux between phases. F_{12}, F_{13}, F_{23} = Interface friction.

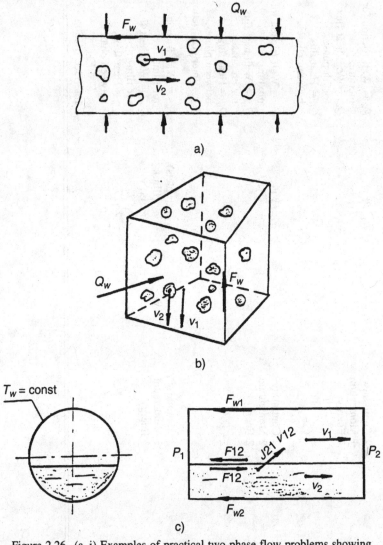

Figure 2.26. (a–i) Examples of practical two-phase flow problems showing relevant parameters.

using an empirical formula. Thus, the two-velocity model is used for this problem.

Figure 2.26d illustrates another case where the two-velocity model may be appropriate. Here a vapor slug flows upward in a vertical pipeline; the temperatures of the phases are equal, but the velocities are different in both magnitude and sign. In this case, only the liquid phase is in contact with

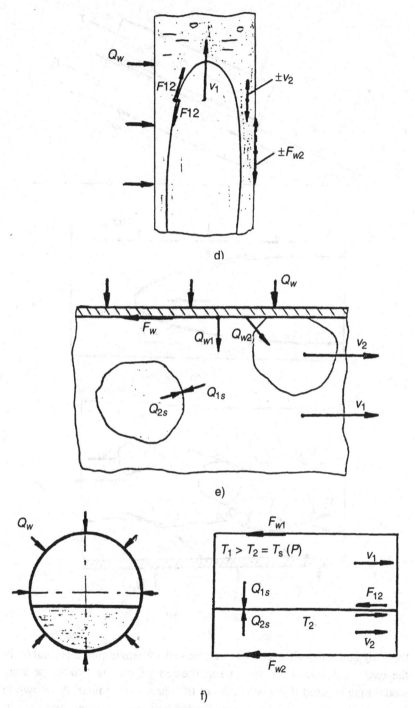

Figure 2.26. *cont.*

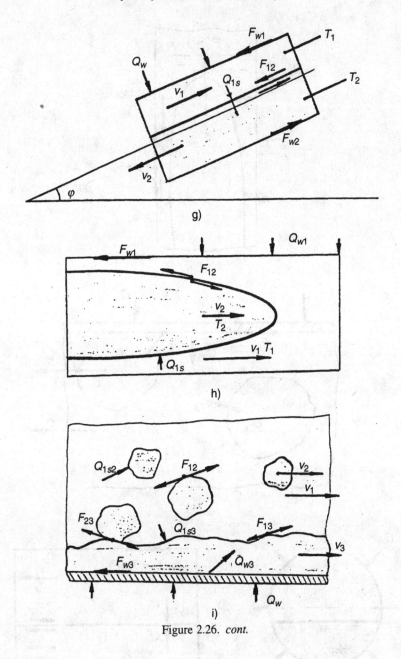

Figure 2.26. *cont.*

the wall and the wall friction is determined by single-phase formulas. In the case of droplet flow within superheated vapor, the two-temperature model may be used if the velocities of the phases are similar. As shown in Figure 2.26e, the wall heat flux is divided between the vapor and the liquid droplets and there is an additional heat transfer between the droplets and the vapor.

The two-velocity, two-temperature model is applicable in a number of situations. In the case of stratified flow with significant amounts of superheat (as opposed to that examined in Section 2.4) both the phase temperatures and velocities differ. Figure 2.26f illustrates this situation. The flow of liquid down an inclined pipeline (Figure 2.26g) may also require the two-velocity, two-temperature model. At sufficient angles of incline the heavy liquid flows down while the lighter displaced vapor flows upward. The initial cooling down of a warm pipeline requires the use of the two-velocity, two-temperature model. The temperatures and velocities of the cold liquid jet entering the pipeline and those of the warm vapor already inside are significantly different. This is shown in Figure 2.26h.

The most complicated example considered here is illustrated in Figure 2.26i. This shows the case of droplet flow within a vapor core that is surrounded by a liquid film at the wall. This is really an example of a three-component model (vapor, droplets, and film). The temperature of the liquid film and the droplets is taken to be the same and there is heat transfer between the wall and the film, the vapor and the film, and the vapor and the droplets. There are friction forces between all three components and between the liquid film and the wall. The solution of this problem is very difficult, requiring not only the simultaneous solution of nine equations but also a very detailed knowledge of the flow regime.

Once the appropriate model is selected, the expressions for the various interactions between the phases must be found. This typically requires the use of empirical data and correlations. The models are then solved using both analytical and numerical techniques.

2.8 Summary

This chapter is concerned with steady-state cryogenic two-phase flow. The general conservation equations for two-phase flow are given. These equations are then used as a starting point to solve problems in the design of cryogenic gasifiers, the stabilization of superconducting magnets, and the description of the geyser effect in fluid transportation systems. The successful solution of these problems requires an understanding of the flow regime present in the problem and the use of some empirically derived relationships. The results of this analysis are in good agreement with experimental data.

The relationship between surface heat transfer and the flow regime was discussed, and regular flows were seen to be the optimal choice to improve the heat transfer. Last, a variety of two-phase flow models was described and related to practical problems for which they are applicable. Later chapters in this book build on this background and extend the analysis to the more difficult area of transient two-phase flow.

3

Transient operating conditions in cryogenic systems with two-phase flow

3.1 Introduction to transient conditions

Practical cryogenic systems are often subjected to transient disturbances. The response of the system to these disturbances can be significantly different than that predicted by steady-state models. In extreme cases, the disturbances can lead to damage to the cryogenic equipment and pose a hazard to personnel. Thus, the response of the system to possible perturbations is an important part of system design.

The transient model of the cryogenic system is based on an understanding of the external disturbances, the geometry of the system, and the applicable physical laws. The model describes how the energy from the disturbance is distributed in the flow, how the perturbation is propagated through the system, and how the flow parameters such as velocity and pressure change in response to the disturbance.

In order to conduct this analysis the nature of the disturbances must be described. In single-phase systems, the disturbances are typically perturbations on the temperature, pressure, or flowrate. In two-phase flow, perturbations affecting the void fraction are also possible. The principal external disturbances that affect a two-phase cryogenic flow then fall into one of the following categories:

1. Change in flowrate (ΔG)
2. Change in system pressure (ΔP)
3. Change in external heat load (ΔQ_w)
4. Change in void fraction ($\Delta \alpha$)

Some examples of these disturbances include the change in the flowrate in a gasification system as users disconnect from it; the pressure change in a liquefaction plant due to the sudden failure of a compressor; and the abrupt increase in heat leak to a magnet cooling channel as a superconducting magnet quenches.

Another issue is determining which portions of the system under study are affected by the disturbance. In some cases the disturbance is restricted to a limited portion of the system. For instance, when outlet valves are closed in a gasification system with a sufficiently large storage tank, the perturbation may be limited to the supply pipelines and not change the conditions in the tank. In many cases, though, the external disturbance will propagate throughout the system, requiring that an accurate technical description of the entire system be included in the model.

The four types of external disturbances listed are not independent. The sudden rise in heat load caused by a magnet quench will in turn cause a sharp rise of the pressure in the cooling channel flow. It is an accurate description of the response to the perturbations that is desired from the transient model.

3.2 Evaluation methods for external disturbances

In order to construct a mathematical model for a cryogenic system responding to external disturbances it is necessary to broadly evaluate the external disturbances in comparison to the system. Are the disturbances large or small? Are they applied smoothly or as a sudden change? The answers to these questions help to determine the form of the resulting model.

In general terms, the impact that an external disturbance has on a system is related to the relative times required for the disturbance to take place and for the system to respond to the disturbance. This second time can also be defined as the time it takes for the disturbance to propagate through the system. If the disturbance occurs much more quickly than the propagation time the impact of the disturbance on the system will tend to be severe. If the perturbation takes place over a long time compared to the propagation time, the impact on the system will tend to be less significant as the system will have more time to adjust to the change.

First, some preliminary definitions are required. The characteristic size of the system (L) is the typical channel or pipeline length in the system. The characteristic time (τ) of the system is defined in terms of the characteristic length and the average steady-state flow velocity ($\bar{v}$):

$$\tau = L/\bar{v}. \tag{3.0}$$

The time of disturbance propagation ($\tau *$) is given by

$$\tau^* = L/c, \tag{3.1}$$

where (c) is the speed of sound in the fluid. Note that this is the shortest time in which a disturbance can propagate throughout the system. Some thermal disturbances may move slower. The characteristic time of the disturbance (T_d) is defined by the disturbance. For a change of flowrate, it may be the actual time it takes to open or close a valve. For changes in external heat load the time is determined by the heat flux, wall thickness, and the wall material. For cases of pressure oscillation the characteristic time is taken to be the period of one oscillation.

3.2.1 Changes in mass flowrate

In large gasification systems the perturbation in flowrate is determined by the rate at which the controlling valve closes. This rate (T_d) is compared to the propagation time of the disturbance (τ^*). If $T_d \ll \tau^*$, then the size of the disturbance will be quite significant. The kinetic energy of the two-phase flow is converted into a pressure wave. Under these conditions, the compressibility of the liquid, gas, and two-phase portions of the flow must be taken into account. The amplitude of the propagating pressure wave may be large enough to cause damage to the gasification system. If, however, the time it takes the valve to close is large compared to the disturbance propagation time $(T_d \gg \tau^*)$, then the system will have time to adjust to the valve closing and the resulting transient disturbance will be comparatively minor. The compressibility of the liquid flow may be neglected in this case.

The determination of τ^* for a gasification system is complicated by the presence of three regions: pure liquid flow, two-phase flow, and pure gas flow. In the two-phase region the speed of sound depends on the void fraction and the propagation of the void fraction is linked to the average velocity of the two-phase flow $(\overline{v_{2ph}})$. The time for the propagation of the disturbance in the two-phase region is thus:

$$\tau^*_{2ph} = \frac{L_{2ph}}{\overline{v_{2ph}}}. \tag{3.2}$$

The propagation time for the gasification system is then the sum of the propagation times for each of the three regions:

$$\tau^* = \frac{L_2}{C_2} + \frac{L_{2ph}}{\overline{v_{2ph}}} + \frac{L_1}{C_1}, \tag{3.3}$$

where C_1 and C_2 are the speed of sound in the gas and liquid flows respectively. Applying Equation (3.3) to the gasification system described in Section 1.1, τ^* is found to be approximately 3 s, which is comparable to the valve closing time during certain emergency conditions. As a result, large pressure spikes may be seen in the system at these times.

3.2.2 Changes in external heat load

The time (T_d) it takes to introduce a change in external heat load into a cooling channel is dependent on the size of the change (ΔQ_w), the mass, and specific heat of the channel wall (M_w, C_{pw}), and a selected rise in wall temperature (ΔT). Conservation of energy then yields:

$$T_d = \frac{(M_w\, C_{pw} DT)}{\Delta Q_w}. \tag{3.4}$$

In the typical operation of the superconducting magnets of the UNK accelerator there will be changes in the heat load due to cycling of the magnet current. In this situation, the wall temperature rise is selected to be approximately 0.1 K and the characteristic time of the disturbance is seen to be between 10^3 and 10^4 s. The time of disturbance propagation (τ^*) for the UNK cooling channels is roughly 1000 s. Thus, the impact of the changing heat load will be fairly minor. For the same system, however, the abnormal condition of a magnet quench dumps so much heat into the cooling channel wall that T_d becomes on the order of 10 to 100 s. As this is now much smaller than the propagation time, the impact of the change is very severe, resulting in sudden pressure rises and violent boiling of the two-phase flow.

3.2.3 Changes in system pressure

In the gasification, transport, and magnet cooling systems analyzed in this work the system pressures change. Examples of these changes include the steady decrease in pressure in the feed tank of a gasification system and the change in pressure in a refrigeration plant due to changes in the operation of the compressor. These changes tend to happen slowly compared to the speed of sound in the two-phase flow and therefore $T_d \gg \tau^*$. If $\bar{v}_i \ll C_{2ph}$, then the change in velocities and pressures can be approximated by

$$\frac{\partial v_i}{\partial t} \sim \frac{v_i}{T_d} \ll v_i \frac{\partial v_i}{\partial z} \sim \frac{v_i^2}{L}; \tag{3.5}$$

and

$$\frac{\partial P}{\partial t} \sim \frac{\Delta P}{T_d} \ll v_i \frac{\partial P}{\partial z} \sim \frac{v_i \Delta P}{L}. \tag{3.6}$$

3.3 Summary

Transient external disturbances are a common event in practical cryogenic systems. The degree to which a disturbance affects the system depends on the speed at which the disturbance occurs and the speed at which the disturbance propagates through the system. The faster the disturbance occurs relative to its propagation time the more likely that the response of the system to the disturbance will be severe. Typical disturbances in two-phase systems are changes in mass flowrate, pressure, heat leak, and void fraction.

Because the disturbances are common and because the response to a transient condition can be very different from the steady-state behavior of the system, modeling the transient response is particularly important. Chapters 4 and 5 examine transient conditions in gasification and magnet-stabilization systems.

4

Transient conditions in
gasification systems

4.1 Results of operating experience

As stated previously, gasification systems are an important application of cryogenic two-phase flows. These systems range in size from small ones having
feed mains only a few meters long to very large ones such as those found in
the Russian launch complex at Baikonur, where the feed mains may exceed
a kilometer in length.

The characteristic transient disturbance in these systems involves changes
to the flowrate, its increase, decrease, or complete termination. These changes
typically result from the connection or disconnection of consuming equipment
at the end of the gasification system. Experience has shown that transient
disturbances to the flowrate result in pressure oscillations in the gasification
system.

Figure 4.1a is a schematic of a large (~1 km in length) nitrogen gasification system. The subcooled liquid nitrogen is fed from a storage tank (1)
through a vacuum-insulated pipeline (2) to a gasifier (3) where heat is applied
to convert the liquid to gas and raise its temperature to the desired value. From
the gasifier the warm gas is transferred via an uninsulated pipeline (4) to a set
of parallel valves (5) connected to the equipment using the gas. The steady-
state temperature distribution of the nitrogen is shown in Figure 4.1b. Note
that the temperature is essentially constant as the liquid moves through the insulated pipeline. There is a step increase in temperature at the gasifier and
(since in this example the temperature of the gas at the outlet of the gasifier
is higher than ambient temperature) a gradual decrease in gas temperature as
it moves through the uninsulated pipeline to the outlet valves.

Experiments were conducted using this gasification system to investigate
the transient pressure oscillations induced by the closing of the valves at the
system outlet. In this study, a steady-state flow of nitrogen gas leaving the
system was established. Once this was done, one or more of the outlet valves

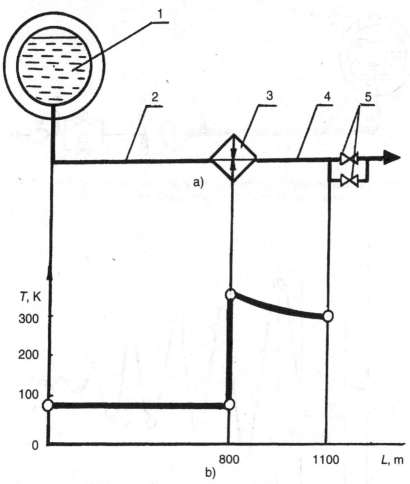

Figure 4.1. (a) Schematic of a large-scale gasification system: (1) liquid supply tank, (2) liquid main, (3) gasifier, (4) gas main, (5) outlet valves. (b) Nitrogen temperature distribution as a function of position in the gasification system: ○ = experimental points.

were closed. The data acquisition system recorded the beginning and end of the valve closing, the changes in the flowrate, and the pressure and temperature changes at various points in the system. Figure 4.2 shows the location of some of the sensors used in the experiment.

An example of the pressure oscillations caused by the closing of one of the outlet valves is shown in Figure 4.3. The pointer indicates the start of the valve closing and τ_{cl} indicates the length of time that the valve required to close. The response of the system to a valve closing is a damped pressure oscillation. Recall that the system under study has two parallel outlet valves. If both valves

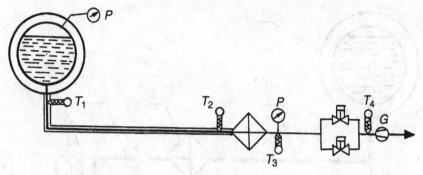

Figure 4.2. Instrumentation schematic of the gasification system experiment. P = pressure transducer; T_1–T_4 = thermometers; G = flowmeter.

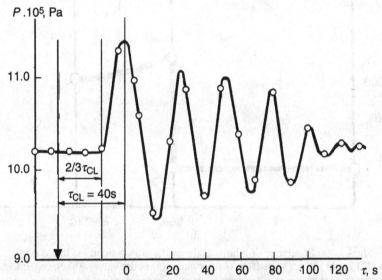

Figure 4.3. Pressure oscillation in the gas main as a result of closing one outlet valve. τ_{cl} = 40 s; $\bigcirc$ = experimental points; $\downarrow$ = start of valve closing.

are closed sequentially, the result is that shown in Figure 4.4. Since the closing of the valves has the effect of superimposing the resulting pressure oscillations, it is entirely possible that the two amplitudes may add together to form a much larger pressure oscillation. To avoid this, the time between valve closings should be chosen to avoid the positive superposition of the pressure oscillations.

An obvious feature of Figures 4.3 and 4.4 is the delay between the start of the valve closing and the first increase in pressure. In these experiments, the initial pressure increase generally occurs about two-thirds of the way through the

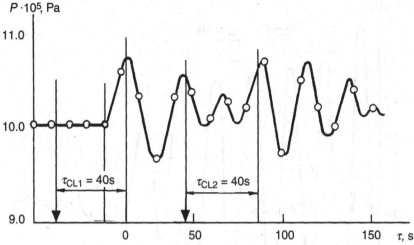

Figure 4.4. Pressure oscillation in the gas main as a result of the sequential closing of two outlet valves. $\tau_{cl} = 40$ s; $\bigcirc$ = experimental points; $\downarrow$ = start of valve closing.

valve closing. This delay is a result of the large size of the gasification system and the fact that in the valves used in the system the flow resistance of the valves increases as the square of the length of travel of the valve plate. Thus, the flow resistance increases much more rapidly near the end of the valve closing time.

The square law relationship between valve closing time and flow resistance defines the nature of the external disturbance in flowrate, $\Delta G(\tau)$. The external disturbance can be related to the valve closing time by

$$\Delta G(\tau) = G_0 - G_0 \left(\frac{\tau}{\tau_{cl}} \right)^2, \tag{4.1}$$

where G_0 is the initial flowrate.

As indicated by Equation (4.1) a shorter valve closing time results in a larger external perturbation ($\Delta G(\tau)$). This in turn results in pressure oscillations that are much larger in amplitude and duration. Figure 4.5 compares the pressure oscillations resulting from two different valve closing speeds. If the oscillations become too big or last too long, significant damage may result to the gasification system. The general rule then is to close the valves slowly enough so that the resulting oscillations are within acceptable limits. One way to reduce the speed of the valves is to mount them so that the closing plate moves against and not with the fluid flow. Thus, the flowing gas acts to slow down rather than speed up the valve closing.

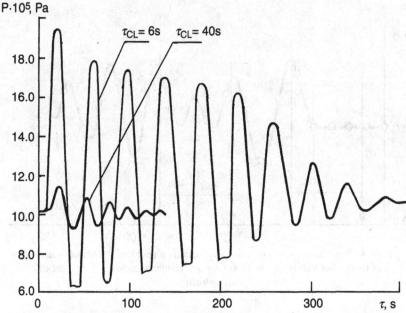

Figure 4.5. Pressure oscillation in the gas main as a function of valve closing speed (τ_{cl}).

There are, of course, certain emergency conditions that require closing the outlet valves much faster than during normal operations. An example is a problem occurring with one of the consuming systems attached to the gasification system. Even here though, the resulting pressure oscillations should be calculated to ensure that appropriate trade-offs have been made and that the cure is not worse than the disease.

The experimental evidence is clear: sudden changes in flowrate in these large gasification systems result in pressure oscillations. The size and duration of these oscillations are closely linked to the rate at which the flow is changed (i.e., to the rate at which the valves are closed). To predict the behavior of the gasification systems under transient disturbances, the physical nature of the oscillations will be described and then a hydrodynamic model developed.

4.2 Physical nature of the oscillations

When the flow is suddenly stopped in the gasification system, the momentum of the gas flowing in the uninsulated gas main is converted to pressure. This causes the pressure at the outlet of the gasifier to be higher than at the inlet,

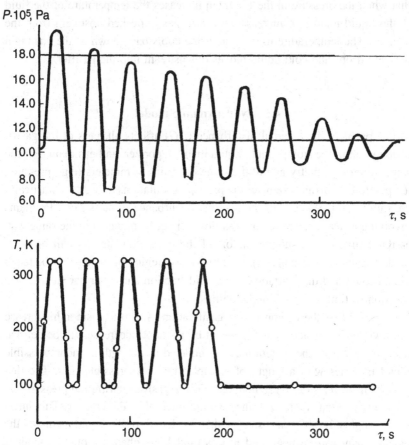

Figure 4.6. Pressure and temperature oscillations in a gasification system as a result of the variation of the flowrate. ○ = experimental points.

forcing the liquid out of the gasifier and back to the storage tank. The large size of this system means that there is a significant amount of liquid between the storage tank and the gasifier. As the pressure at the gasifier falls, this liquid flows back into the still warm gasifier where it vaporizes, resulting in another pressure spike at the gasifier, which in turn starts another oscillation cycle. The oscillations are damped by friction between the wall and the liquid, gas, and two-phase flows. The oscillations halt completely once the liquid stops moving into the gasifier.

Figure 4.6, showing the measured pressure in the gas main downstream of the gasifier and the measured temperature of the fluid upstream of the gasifier as a result of the stoppage of the flowrate, illustrates this process. Notice that when the pressure in the gas main increases the temperature of the fluid in the liquid main also increases as warm gas is pushed upstream from the gasifier. The temperature in the liquid line cools back down as the gas main pressure drops and cold liquid moves downstream into the gasifier.

4.3 Hydrodynamic model

In this problem the external disturbance (ΔG) affects all parts of the gasification system: the storage tank, liquid main, vaporizer, and gas main. For the same system geometry and at the same operating parameters (e.g., pressure, temperature, flowrate) the resulting pressure oscillations may differ drastically depending on the details of the external disturbance in mass flow. Problems involving a steady decrease in mass flow cannot be modeled in the same way as those involving a sudden shutoff of the mass flowrate. As will be seen, modeling the latter requires that the two-phase region of the gasifier be taken into account and that variation of its void fraction and other parameters as a function of time and position be defined.

Consider first the response of a gasification system to a smooth decrease in flowrate. Just exactly what is meant by a smooth decrease will be defined later. In this case, the system may be modeled as a length of incompressible liquid in series with a length of variable (i.e., compressible) gas. The displacement of the liquid–gas boundary in the gasifier during the pressure oscillations is comparable with the packing interval in the gasifier. This interval is roughly 10,000 times smaller than the length of the liquid main and the displacement can be neglected in this model. A schematic of the model is shown in Figure 4.7.

Start by writing the conservation of momentum for the incompressible liquid and the continuity equation for the gas in the gas main:

$$V_l \rho_l \frac{dv_l}{dt} = -F_l - (P_1 - P_0)S_l, \tag{4.1a}$$

where

$$F_l = k_l \frac{\rho_l v_l^2}{2} S_l, \tag{4.1b}$$

$$S_g \rho_g \frac{dv_g}{dt} = \rho_l v_l S_l - \rho_g v_g S_g, \tag{4.1c}$$

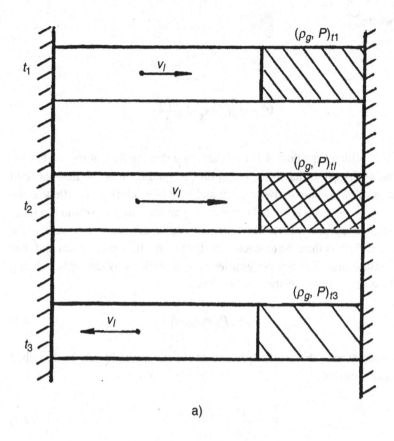

a)

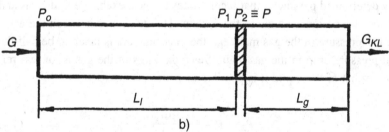

b)

Figure 4.7. (a) Physical model of pressure oscillations in the gasification system. (b) Schematic of simplified model for analysis of transient behavior in the gasification system.

and where

$$V_l = S_l L_l; \quad S_l = \frac{\pi D_l^2}{4}, \quad \text{and}$$

$$V_g = S_g L_g; \quad S_g = \frac{\pi D_g^2}{4}.$$

When writing Equation (4.1c) it is assumed that the flow at the inlet to the gas main is equal to that at the outlet of the liquid main as the mass of fluid in the gasifier is comparatively small and may be neglected. The flow at the outlet of the gas main is related to the closing of the outlet valve and is a function of time: $\rho_g v_g S_g = G(t)$. The time required to set up a steady state in the gasifier is far less than the characteristic time of the flow oscillations and thus the pressure drop through the gasifier may be defined by data taken during steady-state operation of the gasifier. Thus:

$$P_1 - P_2 = f_m(v_l). \tag{4.2}$$

The disturbance to the average pressure in the gas main (P) may be described by a step function:

$$\frac{P}{P_{01}} = \left(\frac{\rho_g}{\rho_{g0}}\right)^\chi, \tag{4.3}$$

where P_{01}, ρ_{g0} are the pressure and density at $t = 0$ and χ is an experimentally determined parameter that incorporates the heat exchange in the gas and the specific design features of the system. The difference between the average gas pressure in the gas main and the inlet pressure is taken to be half the total pressure drop in the gas main. Since the mass of the gas is far less relative to that of the liquid it is possible to write:

$$P_2 - P = \frac{1}{2} k_g \frac{\rho_g v_g^2}{2}. \tag{4.4}$$

In order to solve the system of equations (4.1)–(4.4), the frictional forces are linearized in relation to the initial conditions as

$$F_l = k_l \rho_l v_{l0} \left(v_l - \frac{v_{l0}}{2}\right) S_l, \tag{4.5a}$$

$$F_g = k_g \rho_g v_{g0}\left(v_g - \frac{v_{g0}}{2}\right)S_g, \tag{4.5b}$$

$$f_m(v_l) = f_m(v_{l0}) + k(v_l + v_{l0}), \tag{4.5c}$$

where the subscript 0 refers to the initial conditions. After substitution and double differentiation the following second-order equation is derived:

$$\frac{d^2P}{dt^2} + A_1\frac{dP}{dt} + A_2P + A_3 = 0, \tag{4.6}$$

where

$$A_1 = \frac{1}{L_l}\left(k_l v_{lo} + \frac{1}{2k_g v_{go}}\frac{S_l}{S_g} + \frac{k}{\rho_l}\right),$$

$$A_2 = \frac{S_l P_0 \chi}{L_l V_g S_g},$$

$$A_3 = \frac{P_0 \chi}{V_g \rho_g}\left[A_1\left(\frac{v_{lo}\rho_l S_l}{2} - G_{cl}\right) - \frac{dG_{cl}}{dt} + \frac{S_l}{L_l}\left(\frac{v_{lo}k_l}{2} - f_m(v_{lo}) + P_0\right)\right],$$

G_{cl} = the time dependence of the closing of the outlet valve.

The general analytical solution to Equation (4.6) consists of a sinusoidal oscillating term and an exponential damping term. This solution is given by

$$P(t) = [\exp(-N_1 t)][C_1 \sin\omega t + C_2 \cos\omega t], \tag{4.7a}$$

where

$$N_1 = \frac{A_1}{2} = \frac{1}{2L_l}\left(k_l v_{l0} + \frac{1}{2k_g v_{g0}}\frac{S_l}{S_g} + \frac{k}{\rho_l}\right), \quad \text{and} \tag{4.7b}$$

$$\omega = \frac{\sqrt{4A_2 - A_1^2}}{2}. \tag{4.7c}$$

As $A_1^2 << 4A_2$, we can write:

$$\omega = \sqrt{A_2} = \sqrt{\frac{S_l P_0 \chi}{S_g L_g V_g}},$$ (4.7d)

$$\chi = \frac{S_g L_g V_g \omega^2}{S_e P_0}.$$ (4.7e)

The one-dimensional model created here is, of course, a vast simplification of a real gasification system. Real systems have multiple users connected to parallel gas mains. The geometry of these mains is also complex as they contain bellows, elbows, and transition joints. In addition, there is typically heat transfer between the warm gas in the uninsulated gas main and the ambient environment. These complexities are taken into account by the exponent χ, which Equation (4.7e) shows is empirically determined by the system geometry and the experimentally observed oscillation frequency (ω). For the system studied in this example, χ is found to vary between 0.3 and 0.32. This value of χ is valid for gasification systems with the following parameters: total length between 100 and 1000 m, supply tank pressure between 0.7 and 2 MPa and a flowrate between 5 and 15 kg/s.

Despite its simplicity, the model developed here does explain the oscillation phenomenon and provides important insights into the problem. Equation (4.7d) shows that the oscillation frequency depends on the system geometry, supply pressure, and the coefficient χ but not on the frictional forces between the fluids and the walls. These forces do play an important role in the damping of the oscillations. The stronger the friction forces the more quickly the oscillations will damp out. The time constant of the damping function (N_1) is inversely proportional to the length of the liquid main. This leads to an important yet simple design rule for these gasification systems. That is, the gasifiers should be placed as close to the liquid storage tank as practical. This acts to reduce the length of the liquid main, thus reducing the length of time that the system experiences oscillations after a disturbance in the flowrate. Figure 4.8 shows the effect of the liquid main length on the maximum pressure increase during transient oscillations.

4.3.1 Comparison of model predictions with experimental data

Before comparing the results of this model with experimental data it is instructive to discuss the conditions under which the model will be valid. Recall that in addition to the basic assumptions made at the beginning of the

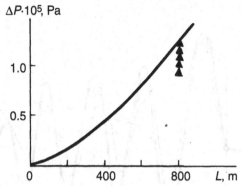

Figure 4.8. Effect of liquid main length on maximum pressure increase in the gasification system. ———— = calculated results; ▲ = experimental points.

model, there is a direct link between theory and experiment by the empirically determined value of χ.

The experiments were conducted using liquid nitrogen and thus the model is valid for LN_2 and also for liquid oxygen, which has similar physical properties. It is not appropriate to use the model for liquid hydrogen and liquid helium without conducting additional experiments with these fluids. The lower molecular weight of LH_2 and LHe means there is significantly less kinetic energy in the moving fluid, which may result in much lower pressure oscillations.

The initial storage tank pressure (P_0) in this experiment ranged between 1.0 and 1.2 MPa. This is far enough away from the critical pressure (3.4 MPa) to allow the gaseous nitrogen to be treated as an ideal gas. Applying this model to systems operating at higher pressures will require that corrections be made to this assumption of ideal gas behavior.

Remember that the flow resistances for both the liquid and gas mains were treated as a lumped resistance for a given value of K_i for each main. Significant changes in the number or types of components (e.g., bellows, elbows) in the mains will necessitate a change in the value of K_i.

The analytical solution of Equation (4.6) is somewhat cumbersome, especially when it is desired to examine the results of closing the two outlet valves sequentially. The results described below come from a numerical solution to Equation (4.6). The numerical technique used is a straightforward finite-difference method. The differential equations are transformed into algebraic equations and the parameters (pressure, velocity, density, etc.) are sequentially determined at a given time step using the known values of the previous time interval.

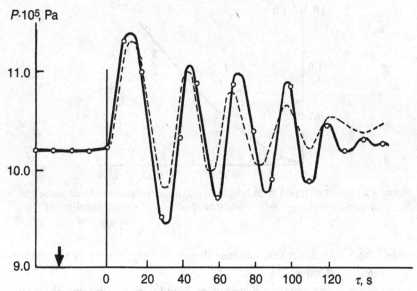

Figure 4.9. Comparison of experimental and calculated results. $\bigcirc$ = experimental points; - - - - - - - = calculated results; $\downarrow$ = start of valve closing.

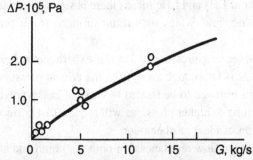

Figure 4.10. Effect of initial flowrate on maximum pressure in the gasification system. ———— = calculated results; $\bigcirc$ = experimental points.

Figure 4.9 compares the predicted pressure oscillation with those found experimentally. In this case, χ was taken to be 0.3. The maximum deviation between the predicted and measured pressure does not exceed 20 percent. The simple one-dimensional model thus permits the prediction of pressure oscillations resulting from the closing of the outlet valves.

The effect of the initial mass flowrate on the maximum amplitude of the pressure oscillation is easily understood. Recall that the oscillations result from the kinetic energy of the fluid being converted into pressure at the end of the

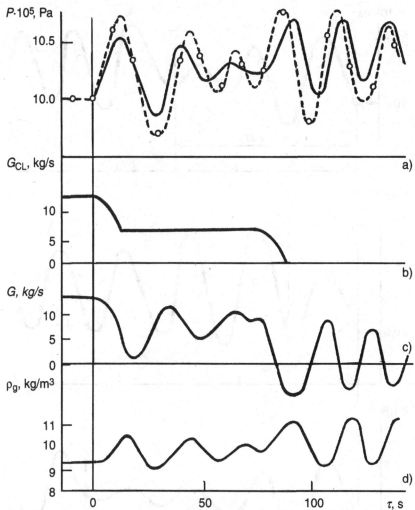

Figure 4.11. Variation of pressure, flowrate, and density in a gasification system due to the sequential closing of two valves. ——— = calculated results; O - - - O - - O = experimental points.

gas main. It is thus reasonable to expect that higher flowrates (i.e., higher kinetic energy) will result in larger pressure oscillations. As shown in Figure 4.10, both the experimental results and model predictions confirm this effect.

Figure 4.11 illustrates the effect of closing the two outlet valves sequentially. The imposed external disturbance (G_{cl}) is shown in Figure 4.11b. After allowing the system to reach steady state, the first valve is closed at time

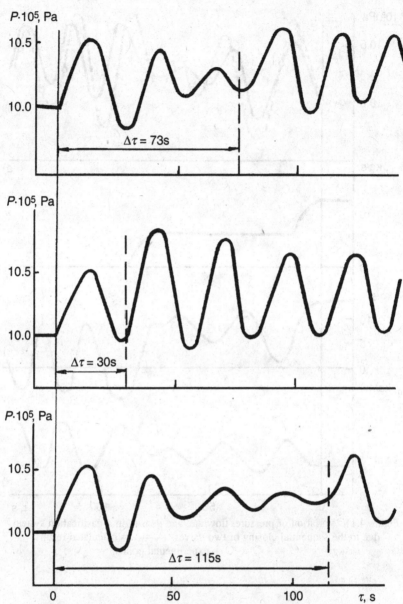

Figure 4.12. Calculated effect of time between sequential closing of two valves ($\Delta\tau$) on the resulting pressure oscillations in the gasification system.

$t = 0$, reducing the mass flowrate G in half. After 70 s the second valve is closed, reducing the mass flowrate to zero. The response of the system pressure, flowrate, and gas density is shown in Figure 4.11 a–c–d. As described before, there are damped oscillations in mass flowrate and pressure as the liq-

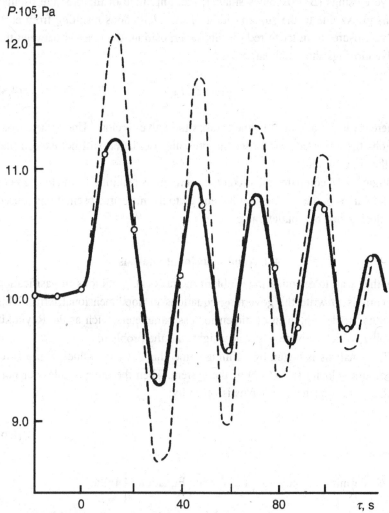

Figure 4.13. Pressure oscillations resulting from the simultaneous closing of two outlet valves compared to the oscillations resulting from the closing of only one outlet valve. ○ = experimental results with the closing of one valve; - - - - - - - = calculated results for the closing of two valves.

uid swings between the storage tank and the gasifier. The gas density also oscillates in phase with the pressure swings.

Numerical simulations of possible operating conditions can be conducted using the hydrodynamic model. The results of these simulations may lead to practical design and operating decisions. For instance, Figure 4.12 shows the effect on the pressure oscillations of the time between the start of valve closing when closing the outlet valves sequentially. Note that as the time between valve closings ($\Delta\tau$) becomes smaller, the amplitude of the pressure oscillations grows due to the superposition of the oscillations resulting from each valve closure. In order to reduce this superposition, the interval between the valve closings should be large, with

$$\Delta\tau \geq 2\,\tau_{\text{cl}}, \tag{4.8}$$

where τ_{cl} is the amount of time it takes the valve to close. Under these conditions the amplitude of the second pressure oscillation will not exceed that of the first oscillation.

Figure 4.13 compares the predicted oscillations resulting from closing both outlet valves of the system simultaneously to the measured oscillations caused by closing just one outlet valve.

4.4 Nondimensional analysis

Another way to examine the problem of transient conditions in gasification systems is to write the governing equations in nondimensional terms. This permits the development of dimensionless parameters, such as the Reynolds number, which provide physical insight into the problem.

This analysis is begun by defining some characteristic values of pressure, time, and velocity (P^*, t^*, v^*) of the system. Then the dimensionless or normalized form of these variables is given by:

$$\bar{P} = \frac{P}{P^*}; \qquad \bar{t} = \frac{t}{t^*}; \qquad \bar{v} = \frac{v}{v^*}. \tag{4.9}$$

These definitions can be used to rewrite Equation (4.1a) as

$$\frac{\rho_l L_l\, v^*\, d\bar{v}}{\rho_l(v^*)^2\, t^*\, d\bar{t}} = \sum_{i=1}^{N} n_i Re_i^{k_i}\, \frac{\bar{v}^2}{2} - \frac{P_1 - P_0}{\rho_l(v^*)^2}. \tag{4.10}$$

Note that the frictional pressure drop is described in terms of the Reynolds number for each of the N components that make up the liquid line. Defining the following dimensionless parameters:

$$\text{Euler number: } Eu_l = \frac{P_1 - P_0}{\rho_l(v^*)^2}. \tag{4.11}$$

$$\text{Strouhal number: } Sh_l = \frac{L_l}{v^* t^*}. \tag{4.12}$$

Equation (4.10) can be written in nondimensional form as

$$Sh_l \frac{d\bar{v}}{d\bar{t}} = \sum_{i=1}^{N} n_i Re_i^{k_i} \frac{\bar{v}^2}{2} - Eu_l. \tag{4.13a}$$

In a similar manner, Equations (4.1b) and (4.1c) may be transformed to

$$Eu_l = \frac{k_l S_l}{v^* \bar{v}}, \tag{4.13b}$$

$$Eu_g = \frac{c_g Sh_g}{v^* v_g}, \tag{4.13c}$$

where c_g is the speed of sound in the gas. The solution to Equation (4.13) will take the general form:

$$\bar{P} = f(\bar{t}, \bar{l}, \bar{v}, Eu_i, Sh_i, Re_i). \tag{4.14}$$

Therefore, the transient response of the gasification system will be determined by the dimensionless time, length, velocity, and the Euler, Reynolds, and Strouhal numbers.

The three dimensionless parameters (*Eu, Re, Sh*) are not just arbitrary collections of terms but, in fact, have definite physical meanings. The Euler number ($Eu_i = \Delta P_i / \rho_i (v^*)^2$) characterizes the transition of flow kinetic energy into the potential energy of pressure. The Strouhal number ($Sh_i = L_i / v^* t^*$) can be thought of as the ratio between the characteristic time of the system and the characteristic time of the disturbance. The familiar Reynolds number ($Re_i = \rho_i v_i D_i / \mu_i$) is the ratio of inertial to viscous forces in the system.

The value of this analysis is illustrated by using the Strouhal number for the liquid main to determine whether pressure oscillations will occur for a gasification system of a given size and valve closing times. In Chapter 3 the presence and significance of pressure oscillations in gasification systems were linked to the relative size of the disturbance time (T_D) and the response time of the system (τ^*). The Strouhal number for the liquid main defines this ratio. The Strouhal number for the liquid main is

$$Sh_l = \frac{L_l}{(c_l + v_l)\tau_{cl}}, \tag{4.15}$$

where c_l is the speed of sound in the liquid main and τ_{cl} is the time it takes the valve to close. Equation (4.15) is seen to be the ratio of the time it takes the system to respond to changes to the time it takes the valve to close. Experience with large commercial gasification systems has shown that the response of the system to transient conditions may be divided into three classes. If $Sh_l < 0.01$, then $T_D >> \tau^*$ and no significant pressure oscillations will occur when the outlet valve is closed. If $0.1 \geq Sh_l \geq 0.125 - 0.01$, then oscillations will occur but the compressibility of the liquid may be neglected and the hydrodynamic model developed in Section 4.3 may be used provided the previously described assumptions of the model are true. If, however, $Sh_l > 0.1$, then $T_D << \tau^*$ and the speed of the valve closing is much faster than the response time of the system and the compressibility of the gas must be considered in the analysis of the problem. This situation usually occurs when the outlet valves are closed suddenly in response to some emergency conditions. This problem will be considered in the next section.

Thus, the nondimensional analysis results in a simple parameter, the Strouhal number, which allows the gasification system designers to easily predict the presence of pressure oscillations and the appropriate model to use in analyzing them.

4.5 Analysis of high-speed transients

When the Strouhal number for the liquid main (Sh_l) is greater than 0.1, then the speed of the valve closing is much faster than the characteristic speed of the system. Under these conditions, the model developed in Section 4.3 cannot be used and a more complicated model including compressibility of the liquid and of the two-phase mixture in the gasifier must be used.

This model is shown schematically in Figure 4.14, where L represents the liquid main, D represents the portion of the gasifier containing a two-phase mixture, I represents the portion of the gasifier containing single-phase gas, and G represents the gas main. For each of these regions, the three conservation equations of mass, momentum, and energy must be written. Since the flow in the liquid main is isothermal, the energy equation in this section will be identically zero.

The conservation equations may be written in quasi-divergent form as:

$$\frac{\partial \overline{U}}{\partial t} + \partial \overline{F(\overline{U})} \backslash \partial x = f(\overline{U}), \tag{4.16}$$

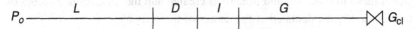

Figure 4.14. Diagram of high-speed transient model: L = the liquid main; D = portion of the gasifier containing two-phase flow; I = portion of gasifier containing single-phase gas; G = gas main.

where the vectors are defined as

$$\bar{U} = \begin{bmatrix} \rho \\ m \\ e \end{bmatrix}; \qquad \overline{F(\bar{U})} = \begin{bmatrix} m \\ \dfrac{m^2}{\rho} + P \\ (e + P)\dfrac{m}{\rho} \end{bmatrix}'; \qquad f(\bar{U}) = \begin{bmatrix} 0 \\ F_w \\ Q \end{bmatrix};$$

(4.17)

where

$$e = \rho\left(u + \frac{v^2}{2}\right),$$

u = the specific internal energy,

m = ρv,

Q = the external heat input,

F_w = the wall friction force.

An extensive literature exists describing the solution of boundary-value problems such as this using numerical methods. These techniques include the method of characteristics, etc. In these methods, the differential equations are replaced by algebraic ones and the solutions are points of a set referred to as a network.

Good results have been achieved in using the double interval technique of Lax–Wendroff to solve equations of the form given by Equation (4.16). These equations are written for a space with unit volume. If a region is enclosed between x_1 and x_2 then the full amount of a value per unit area is equal to the component integral. The algebraic difference equations used in the numerical method have second-order accuracy and the method is conditionally stable for hyperbolic equations such as Equation (4.16). The advantage of the Lax–Wendroff method is that the algebraic difference equations are also written in the form of conservation equations. Thus, values that are conserved in the differential equations are exactly conserved in the difference equations.

The principle of the Lax–Wendroff technique is as follows. A two-dimensional mesh in space (x) and time (t) is created and the intermediate values of the vector $\overline{U}$ are calculated at the center of the cells by

$$\overline{U}_{j+1/2}^{n+1/2} = \frac{1}{2}(\overline{U}_{j+1}^{n} + \overline{U}_{j}^{n}) - \frac{\Delta t}{2\Delta x}(\overline{F}_{j+1}^{n} - \overline{F}_{j}^{n}). \qquad (4.18)$$

where $\overline{F}_{j}^{n} = \overline{F}(\overline{U}_{j}^{n})$ and j and i are indexes denoting positions in space, x, and time, t, respectively.

The final results are then obtained by

$$\overline{U}_{j}^{n+1} = \overline{U}_{j}^{n} - \frac{\Delta t}{\Delta x}(\overline{F}_{j+1/2}^{n+1/2} - \overline{F}_{j-1/2}^{n+1/2}). \qquad (4.19)$$

The main parameters of ρ, m, e are thus found at each interval of time and position.

This technique is stable if

$$\Delta t \le \frac{\Delta x}{v + c}, \qquad (4.20)$$

where c is the velocity of sound in the region under study. This criterion results in a very fine time mesh that increases the computation time and computer memory requirements for this technique.

The right-hand cell appears according to the Lax–Wendroff diagram. It describes the valve operation ($G_{cl} = f(t)$). At the liquid section it should be joined to the conservation equations of the two-phase flow.

As in Chapter 2 the equilibrium model ($v_1 = v_2 = v$) will be used to describe the two-phase region of the gasifier. Thus from Equation (2.8) we can write:

$$\frac{\partial(\rho_1^0 \alpha_1)}{\partial t} + \frac{\partial(\rho_1^0 \alpha_1 v)}{\partial x} = J_{21}, \qquad (4.21a)$$

$$\frac{\partial(\rho_2^0 \alpha_2)}{\partial t} + \frac{\partial(\rho_2^0 \alpha_2 v)}{\partial x} = -J_{21}, \qquad (4.21b)$$

$$(\rho_1^0 \alpha_1 + \rho_2^0 \alpha_2)\left(\frac{\partial(v)}{\partial t} + \frac{v\partial(v)}{\partial x}\right) = -\frac{\partial P}{\partial x} - F_w, \qquad (4.21c)$$

where

$$\alpha_1 + \alpha_2 = 1,$$
$$\rho_2^0 = \text{constant},$$

$$\rho_1^0 = \frac{P}{RT_s(P)},$$

$$J_{21} = \frac{Q_w}{\zeta}.$$

The wall friction force (F_w) is found via the Lockhart–Martinelli correction, Equation (2.15).

The characteristic time for the pressure and velocity of the two-phase mixture to respond to changes in the system is given by

$$\Delta t = \frac{\Delta x}{c_{2ph}} \sim \frac{10^{-1}}{10} = 10^{-2} \text{ s,} \qquad (4.22)$$

where c_{2ph} is the speed of sound in the two-phase mixture. This time is far less than the valve closing time ($\tau_{cl} \sim 6 - 40$ s) and thus the pressure and velocity of the two-phase region may be approximated as independent of time. The characteristic time for the change of the void fraction (α_1) is

$$\Delta t = \frac{\Delta x}{v_{2ph}} \sim \frac{10^{-1}}{10^{-1}} = 1 \text{ s.} \qquad (4.23)$$

This is much slower and the time dependence of the void fraction may be derived from Equation (4.21b):

$$\frac{\partial \alpha_1}{\partial t} = -v\frac{\partial \alpha_1}{\partial x} + (1 - \alpha_1)\frac{\partial v}{\partial x} + \frac{Q_w}{\rho_2^0 \zeta}. \qquad (4.24)$$

This allows $\alpha_1(t)$ to be calculated in each time step based on the previous time step.

Since in the two-phase region, $\partial v/\partial t \sim 0$, Equation (4.21) may be combined to form:

$$\alpha_1 v \left(\frac{\partial \rho_1^0}{\partial P}\right)\frac{\partial P}{\partial x} + (\rho_1^0 \alpha_1 + \rho_2^0 \alpha_2)\frac{\partial v}{\partial x}$$

$$\qquad (4.25a)$$

$$= (\rho_2^0 - \rho_1^0)v\left(\frac{\partial \alpha_1}{\partial x}\right),$$

$$\frac{\partial P}{\partial x} + (\rho_1^0 \alpha_1 + \rho_2^0 \alpha_2)v\frac{\partial v}{\partial x} = -F_w, \qquad (4.25b)$$

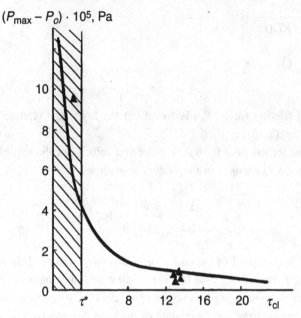

Figure 4.15. Influence of valve closing time on pressure amplitude. ——— = cal-culated results; ▲▲ = experimental results; τ^* = characteristic time of system. Shaded area denotes region where the valve closing time is less than the characteristic time of the system.

where

$$\frac{\partial \rho_1^0}{\partial P} = \frac{\partial \left(\dfrac{P}{RT_s(P)} \right)}{\partial P}.$$

These two equations can be solved numerically to find the two unknowns $\partial P/\partial x$, $\partial v/\partial x$ and thus new values of P and v can be computed based on the left-hand boundary values of P_1^n, v_1^n.

As can be seen, solving the problem of high-speed transients in gasification systems is a complicated two-dimensional problem requiring significant computer memory and processor time. Fortunately, the nondimensional analysis described in Section 4.4 allows the definition of the operating region in which these high-speed transients occur. Once this region is defined, the gasification system should not be operated in this region except under the most extreme emergency conditions. Figure 4.15 illustrates the potential hazard of operating a gasification system with high-speed transients. Note that as the value of the valve closing time (τ_{cl}) becomes less than the characteristic time of the system (τ^*), the maximum amplitude of the pressure oscillation increases dramatically. The figure also compares the predictions of the numerical model to the experimentally observed values.

4.6 Practical considerations in the design of gasification systems

The results of the experimental work and analysis described in this chapter lead to some basic guidelines that should be considered when designing large-scale cryogenic gasification systems.

1. The length of the liquid main is seen to have a great impact on the amplitude and duration of the pressure oscillations. In general, the gasifier should be placed as close as possible to the liquid storage tank. This reduces the length of the liquid main, which in turn reduces the amplitude and duration of the pressure oscillation resulting from changes in the flowrate.

2. For a system that has a number of parallel outlet valves, the time interval between the sequential closings of the valves should be chosen so that the resulting pressure oscillations do not constructively interfere, resulting in a larger total pressure spike. For two valves of the same diameter, the time between the start of each valve closing should be greater than or equal to twice the valve closing time to prevent this positive interference.

3. As described in Section 2.3, the gasifier should be designed to produce the optimum dispersed flow that has the maximum surface area of phase transition. This can be done by using the staggered fin assembly described in Chapter 2. Gasifiers designed to generate these optimum two-phase flows will be the most efficient although the resulting increased pressure drop of such designs should be considered.

4. The system should always be designed so that it does not operate in a region that generates high-speed transients. That is, the valve closing time (τ_{cl}) should be much much longer than the characteristic time of the system (τ^*). Alternatively this criterion is met if the nondimensional Strouhal number of the liquid main (Sh_l) is less than 0.01. If these conditions are not met, the pressure oscillations resulting from flowrate changes will be very high in amplitude and may cause damage to the system.

5. One technique for increasing the closing time of the valves is to install the outlet valves in such a way that the fluid flow impedes rather than assists the valve closing.

6. When modeling the behavior of a gasification system, the Strouhal number of the liquid main may also be used to determine the appropriate model

for the analysis. If $0.1 \geq Sh_l \geq 0.0125 - 0.01$, then the compressibility of the liquid may be neglected and the one-dimensional, $P(\tau)$, hydrodynamic model developed in Section 4.3 may be used. If, however, $Sh_l > 0.1$, then the compressibility of the gas must be considered and the more complicated two-dimensional, $P(x,\tau)$, must be employed.

Remember that the analysis of the gasification systems described in this chapter had an empirical component in the term of χ. Thus the results of this chapter can only be reliably applied to similar systems. For example, the working fluid should be either nitrogen or oxygen. However, the techniques developed in this chapter can be applied to other gasification systems as long as some experimental work is done to provide the appropriate empirical factors.

4.7 Summary

The subject of this chapter has been the transient behavior of cryogenic gasification systems. The physical response of these systems to the sudden change of flowrate has been described and appropriate analytical models developed. As discussed in Chapter 3, the principal factor is the relative speeds of the external disturbance and the system response. A nondimensional analysis has produced the Strouhal number, which indicates both the impact of the disturbance on the system and the appropriate model to be used in the analysis.

The results of this work show that it is possible to generate a useful predictive model (albeit one that requires some empirical data) of a complicated large-scale gasification system. This work has been used to produce an efficient, reliable gasification system for the Russian Energia–Buran launch complex.

5

Transient conditions in
magnet-stabilization channels

5.1 Modeling of transients resulting from variable heat loads

The advantages of using two-phase helium flows to stabilize superconducting magnets were described in Section 2.4. This chapter is concerned with the transient response of the magnet cooling channels to external disturbances. The most relevant external disturbance affecting these cooling systems is a change in the heat load to the cooling channels (ΔQ). The examples in this chapter will be drawn from the area of magnets designed for large accelerators used for research in high-energy physics. The magnet systems used in these accelerators share a number of demanding characteristics. The scale of the systems tends to be very large, with characteristic lengths on the order of kilometers. The magnets operate at relatively high fields (4–9 T) and thus store a significant amount of energy. Additionally, the size and cost of these systems require that they operate reliably for long periods under a variety of changing conditions.

The heat load to the cooling channels of these magnets may be divided into several components. First, there is the steady-state heat load that results from heat leaking into the cryostat from the external environment. Next, there is a periodically changing heat load that results from alternating current (ac) losses in the superconductor as the magnet current is changed, from resistive losses in nonsuperconducting splices within the magnet circuit, and from radiation emitted by the particle beam traveling through the beam tube at the center of the magnet. Last, in the rare (it is hoped) case of a magnet quench, a large amount of heat is quickly deposited into the helium in the magnet cooling channels.

The cross section of the dipole magnets designed for use in the planned UNK accelerator in Russia is shown in Figure 1.9. Here the heat load is deposited into a supercritical helium flow ($P = 3 \times 10^5$ Pa, $T = 4.5$ K) flowing through the annular space (73 mm $\times$ 80 mm) that surrounds the beam tube. A portion of this single-phase flow is fed into crescent-shaped channels where it is cooled

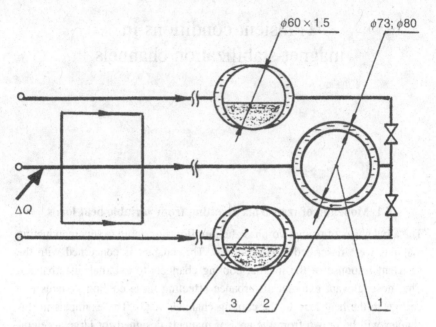

Figure 5.1. Schematic of UNK Magnet Cooling System: (1) beam tube, (2) central
single-phase cooling channel, (3) single-phase bypass channel, (4) two-phase
flow channel.

by a counterflow of two-phase helium. The recooled single-phase flow is then
recombined with the helium in the annular space. This separating and recom-
bining of flows occurs every 6 m. Figure 5.1 shows this cooling scheme. The
cryogenics system for the UNK machine is divided into arms, each cooling 650
m of magnets. This is shown in Figure 1.8. Note that the output of the two-
phase cooling channels is a saturated helium bath (item 7 in Figure 1.8). The
varying heat load due to ac losses and beam radiation is shown in Figure 5.2.
The effect of this disturbance on the two-phase flow may be to cause an irregular flow into the saturated bath, which may dry out or overheat. Additionally,
over the long operation period envisioned for the accelerator (the system is designed to operate continuously for up to 3 months at a time) the changing heat
load may result in a lowering of the liquid level in the two-phase pipes, resulting in superheating of the vapor and potentially quenching of the magnets.

The goal of the modeling, then, is to investigate the response of system parameters (e.g., velocity and pressure) to the periodically varying heat load to
determine the possibility of significant vapor superheat or the drying out or
overfilling of the saturated helium bath. This model is not appropriate for
studying the much rarer (and much more violent) magnet quench.

As a first step, the system shown in Figures 5.1 and 1.8 may be modeled

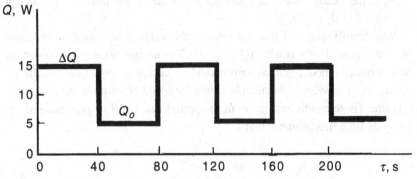

Figure 5.2. Expected periodic heat load deposited in the UNK magnets.

by a simple pipe ($L = 650$ m, $OD = 60$ mm) containing two-phase flow. The boundary conditions on this pipe correspond to the actual physical conditions on the cooling channel such as the throttling of the flow at the inlet to the channel.

In Chapter 2 the analysis of the steady-state behavior of the two-phase flow in the UNK magnet cooling channels showed that the flow was stratified and that there was a significant amount of slip. That is, the velocity of the vapor moving in the upper part of the channel was faster than the velocity of the liquid moving in the lower part. This required the use of the two-velocity model. For the transient case though, it is permissible to approximate the velocities of the two components as being equal. This will allow the problem to be described using the equilibrium model ($v_1 = v_2 = v$, $T_1 = T_2$). This simplifies the problem significantly as it is already a two-dimensional problem with variations in both time and space.

Based on these assumptions we may write the time-dependent continuity equations for each phase and the total flow momentum equation based on Equation (2.5):

$$\frac{\partial(\rho_1^0 \alpha_1)}{\partial t} + \frac{\partial(\rho_1^0 \alpha_1 v)}{\partial z} = J_{2,1}, \tag{5.0a}$$

$$\frac{\partial(\rho_2^0 \alpha_2)}{\partial t} + \frac{\partial(\rho_2^0 \alpha_2 v)}{\partial z} = -J_{2,1}, \tag{5.0b}$$

$$\frac{\partial(\rho_m v)}{\partial t} + \frac{\partial(\rho_m v^2)}{\partial z} = -\frac{\partial(P)}{\partial z} + F_w, \tag{5.0c}$$

where $\rho_m = \rho_1^0 \alpha_1 + \rho_2^0 \alpha_2$.

As in the steady-state case analyzed in Chapter 2, the phase transition can be defined as $J_{2,1} = Q_w/\zeta$.

Note from Figure 5.2 that the ratio of the varying heat load to the static heat load is relatively small ($\Delta Q/Q_0 \sim 3$). This means that there is no sudden large explosive heat input, as there would be during a magnet quench. In addition, the flow velocity is much less than the speed of sound in the two-phase mixture. These conditions permit us to approximate the flow pressure and velocity as time independent, that is:

$$\frac{\partial P}{\partial t} \sim \frac{\partial v}{\partial t} \sim 0. \tag{5.1}$$

As will be seen, however, the void fraction (α) is a function of time. The assumption that the pressure is time independent also allows the liquid to be treated as an incompressible fluid and the vapor as an ideal gas. Thus:

$$\rho_1^0 = P/RT_s(P); \tag{5.2a}$$

$$\rho_2^0 = \text{constant}. \tag{5.2b}$$

Furthermore, calculations have shown that for relatively small changes in the heat load ($\Delta Q/Q_0 < 100$), the variation in system pressure ($\Delta P/P_0$) is not very significant. Thus, we may also approximate ρ_1^0 as constant. Note that these assumptions would not be valid in the case of a quench.

Now continue the analysis by inserting the definition of ρ_m in Equation (5.0c) and expanding the left-hand side of the equation to get:

$$\frac{\partial((\rho_1^0\alpha_1)v)}{\partial t} + \frac{\partial((\rho_2^0\alpha_2)v)}{\partial t} + \frac{\partial((\rho_1^0\alpha_1 v)v)}{\partial z} + \frac{\partial((\rho_2^0\alpha_2 v)v)}{\partial t}. \tag{5.3}$$

Now explicitly carrying out the differentiation of Equation (5.3) yields:

$$v\frac{\partial(\rho_1^0\alpha_1)}{\partial t} + \rho_1^0\alpha_1\frac{\partial v}{\partial t} + v\frac{\partial(\rho_2^0\alpha_2)}{\partial t} + \rho_2^0\alpha_2\frac{\partial v}{\partial t},$$

$$+ v\frac{\partial(\rho_1^0\alpha_1 v)}{\partial z} + \rho_1^0\alpha_1 v\frac{\partial v}{\partial t} + v\frac{\partial(\rho_2^0\alpha_2 v)}{\partial z} + \rho_2^0\alpha_2 v\frac{\partial v}{\partial t}. \tag{5.4}$$

Recall that $\partial v/\partial t = 0$ and that from Equation (5.0a,b)

$$\frac{\partial(\rho^0{}_i\alpha_i)}{\partial t} + \frac{\partial(\rho^0_i\alpha_i v)}{\partial z} = (-1)^{i+1} Q/\zeta. \tag{5.5}$$

We can write Equation (5.4) as

$$(vQ/\zeta - vQ/\zeta) + (\rho^0_1\alpha_1 + \rho^0_2\alpha_2)v\frac{\partial v}{\partial t}. \tag{5.6}$$

This reduces to (using the definition of ρ_m) $\rho_m v \, \partial v/\partial z$. Thus Equation (5.0c) may be simplified to

$$\rho_m v\frac{\partial v}{\partial z} = -\frac{\partial P}{\partial z} + F_w. \tag{5.7}$$

Since we are taking ρ^0_1 and ρ^0_2 as constant, Equation (5.0 a,b) may be written as

$$\frac{\partial\alpha_1}{\partial t} + \frac{1}{\rho^0_1}\frac{\partial(\rho^0_1\alpha_1 v)}{\partial z} = \frac{1}{\rho^0_1}\frac{Q}{\zeta}, \tag{5.8a}$$

$$\frac{-\partial\alpha_1}{\partial t} + \frac{1}{\rho^0_2}\frac{\partial(\rho^0_2\alpha_2 v)}{\partial z} = -\frac{1}{\rho^0_2}\frac{Q}{\zeta}. \tag{5.8b}$$

Adding these last two equations together and simplifying the left-hand side produces:

$$\frac{\partial v}{\partial z} = \frac{Q}{\zeta}\left(\frac{1}{\rho^0_1} - \frac{1}{\rho^0_2}\right). \tag{5.9}$$

Differentiating Equation (5.0a) we get

$$\rho^0_1\frac{\partial\alpha_1}{\partial t} + \rho^0_1 v\frac{\partial\alpha_1}{\partial z} + \rho^0_1\alpha_1\frac{\partial v}{\partial z} = \frac{Q}{\zeta}. \tag{5.10}$$

Substituting Equation (5.9) into Equation (5.10) yields:

$$\frac{\partial\alpha_1}{\partial t} = -v\frac{\partial\alpha_1}{\partial z} + \frac{Q}{\zeta}\left(\frac{1 - \alpha_1}{\rho^0_1} + \frac{\alpha_1}{\rho^0_2}\right). \tag{5.11}$$

These manipulations have produced a set of three equations:

$$\frac{\partial \alpha_1}{\partial t} = -v\frac{\partial \alpha_1}{\partial z} + \frac{Q}{\zeta}\left(\frac{1-\alpha_1}{\rho_1^0} + \frac{\alpha_1}{\rho_2^0}\right), \tag{5.12a}$$

$$\frac{\partial v}{\partial z} = \frac{Q}{\zeta}\left(\frac{1}{\rho_1^0} - \frac{1}{\rho_2^0}\right), \tag{5.12b}$$

$$\frac{\partial P}{\partial z} = -(\rho_1^0 \alpha_1 + \rho_2^0 \alpha_2)v\frac{\partial v}{\partial z} + F_w, \tag{5.12c}$$

with three unknowns $(\alpha_1(t, z), v\,(t, z), P(t, z))$.

Thus, by making the assumptions of equilibrium flow, a relatively small change in pressure, and a time-independent velocity, the transient response of the magnet cooling channel to a variable heat load is described by a fairly simple set of partial differential equations.

Looking at Equation (5.12) it is obvious that as the heat load (Q) is a function of time both the velocity and pressure must be time dependent. This appears to contradict the initial assumption that the pressure and velocity were time independent. In fact, the solution to Equation (5.12) will show that the change of velocity and pressure with time is relatively small and that the initial approximation is still valid.

The boundary conditions for Equation (5.12) are based on the physical conditions of the problem. Two-phase flow is formed immediately after the throttling valve at the inlet of the cooling channel. One of the boundary conditions, therefore, is that the enthalpy at the channel inlet is constant. This enthalpy is the weighted sum of the component enthalpies at the channel entrance:

$$\rho_1^0 \alpha_1 I_1 + \rho_2^0 \alpha_2 I_2 = \text{constant}, \tag{5.13}$$

where $I_i = u_i + P/\rho_1^0$, $i = 1,2$, u_i being the internal energy of the ith component at the cooling channel inlet. As this is a two-phase mixture the component enthalpies are related by the heat of vaporization ($I_1 - I_2 = \zeta$). The boundary conditions at the channel inlet may thus be written:

$$\rho_1^0 \alpha_1 u_1 + \rho_2^0 \alpha_2 u_2 = \text{constant}, \tag{5.14}$$

where $u_2 = Cv_2 T_s$ and $u_1 = u_2 - P(1/\rho_1^0 - 1/\rho_2^0) + \zeta$.

The system is controlled so that the mass flowrate G is constant at the inlet of the cooling channel. So a second boundary condition is

$$G_{\text{inlet}} = \text{constant.} \tag{5.15}$$

The third and final boundary condition is given by the fact that the cooling channel feeds into a large saturated bath. The volume of this bath is significant in the UNK machine (1.96 m^3) Due to this size, the pressure of this bath may be taken as constant even under the transient conditions. As the bath pressure is also the outlet pressure of the cooling channel, it is correct to say that:

$$P_{\text{outlet}} = \text{constant.} \tag{5.16}$$

The initial conditions of the problem are given by the steady-state values of α_1, v, and P that were present before the external disturbance was applied. This problem was solved in Section 2.4.

Equation (5.15) is solved numerically via a standard two-dimensional finite-difference technique. In this method, the differential equations are replaced with algebraic difference equations. For example, (5.15a) is rewritten as

$$\frac{(\alpha_1)_i^{n+1} - (\alpha_1)_i^n}{\Delta t} = \frac{Q^n}{\zeta}\left[\frac{1 - (\alpha_1)_i^n}{\rho_1^0} + \frac{(\alpha_1)_i^n}{\rho_2^0}\right]$$

$$-v_i^n\left(\frac{(\alpha_1)_i^n - (\alpha_1)_{i-1}^n}{\Delta z}\right), \tag{5.17}$$

where i and n are cell indexes in space and time respectively.

In this problem, the space and time step sizes are related to each other by the mean velocity of the flow:

$$\Delta t \le \frac{\Delta z}{\bar{v}}. \tag{5.18}$$

5.2 Results of the analysis

The heat load to the magnet cooling channel, while varying, is periodic (see Figure 5.2), and it is reasonable to expect that after an initial transient response the system would settle down to a new quasi steady state driven by the periodically varying heat load. In fact, the solutions to Equation (5.12) show that this is exactly what occurs. Starting with a constant heat load, such as would be seen with no magnet ramping or beam present, and then adding the varying heat load, the flow parameters are seen to oscillate for a period of time and then reach a quasi steady state. The characteristic time of the transient period between the two steady states may be estimated from the average velocity of the two-phase flow and the length of the magnet cooling channel. In the case of the UNK machine, this time becomes,

$$L/\bar{v} = 650 \ m/0.4 \ m/s \sim 1600 \ s. \tag{5.19}$$

5.2.1 Variation in pressure

The response of the pressure in the two-phase magnet cooling channel to the varying heat load is shown in Figure 5.3a. These results come from the solution of Equation (5.12). As expected, the pressure varies as both a function of time and position. Note that the change in pressure is fairly small ($\Delta P/P_0 \sim$ 5%) and thus the approximations made about constant density and time-independent pressure in the problem are valid.

The pressure as a function of time at the channel inlet is shown in Figure 5.3b. The pressure oscillates with a frequency that is close to the frequency of the varying heat load (see Figure 5.2). In addition to this oscillation there is a slower decrease in pressure seen in Figure 5.3b as the system moves to a new quasi steady state. In this quasi steady state the pressure will oscillate but the time-averaged pressure will remain constant. The time for this new quasi steady state to be reached is seen to be somewhat larger than 1000 s.

Due to the wall friction, the pressure decreases as a function of position along the pipe for a given time as shown in Figure 5.3c.

5.2.2 Variation in velocity

Figure 5.4a–c shows the predicted change in velocity due to the varying heat load. The increased pressure at the inlet of the channel results in an increase in the flow velocity as the channel outlet pressure is held constant by the saturated bath into which the channel empties. The velocity increases along the length of the channel as the two-phase flow absorbs additional heat from the magnet. This increase is shown in Figure 5.4b. The time dependence of the velocity is shown in Figure 5.4c. Again, the period of oscillation is very similar to that of the frequency of the disturbance ($\Delta Q(t)$).

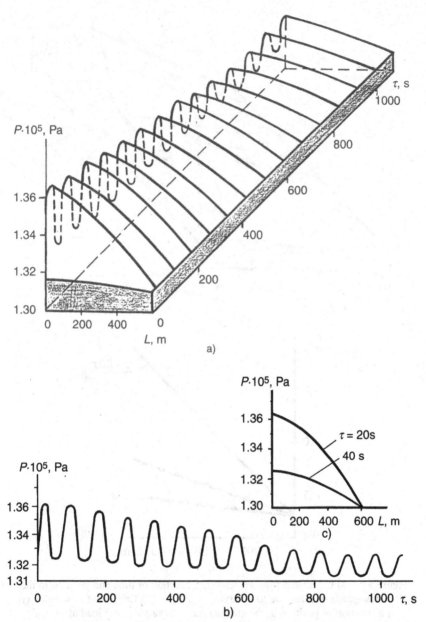

Figure 5.3. (a) Calculated pressure oscillations as a function of time and position in the UNK two-phase channel due to a varying heat load. (b) Calculated pressure oscillations at the inlet to the UNK two-phase channel as a function of time. (c) Pressure distribution in the UNK two-phase channel as a function of position in the channel.

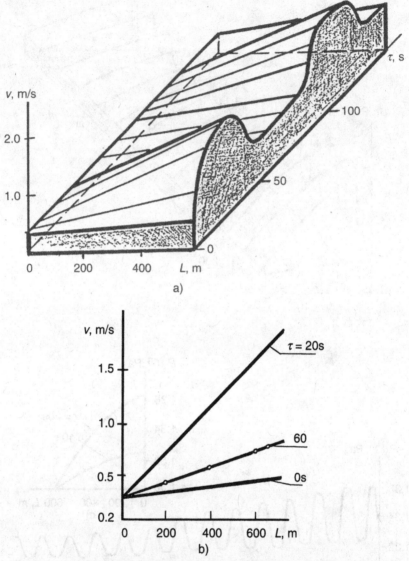

Figure 5.4. (a) Calculated flow velocity as a function of time and position in the UNK two-phase channel due to a varying heat load. (b) Calculated flow velocity as a function of position in the channel due to a varying heat load at several different times.

5.2.3 Variation in void fraction

Recall from the steady-state analysis of this problem done in Section 2.4 that the two-phase flow in the UNK cooling channels is stratified and that if the liquid level in the channel drops low enough, superheating of the vapor may

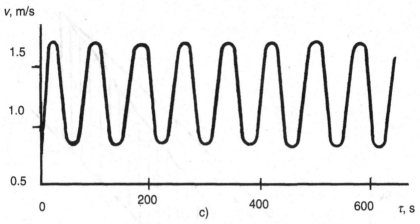

Figure 5.4. (c) Calculated time variation of flow velocity at UNK two-phase channel outlet.

occur, resulting in quenching of the magnets. Because the void fraction (α_1) in a stratified flow is inversely proportional to the liquid level, the response of the void fraction to the varying heat load is particularly important.

The predicted response is shown in Figure 5.5a–c. The void fraction increases slowly with both time and position along the channel. There is a slight oscillation (at near the frequency of the varying heat load) of the changing void fraction. The void fraction is approaching its new quasi-steady value at a time somewhat greater than 1000 s.

Large void fractions like those seen in Figure 5.5c, at times approaching 1000 s, may result in magnet quenching. To prevent this, a higher initial mass flowrate should be used so that the resulting quasi-steady void fraction caused by the varying heat load is lower and won't affect magnet stabilization.

5.2.4 Variation in liquid flowrate

To ensure proper operation, the saturated bath into which the magnet cooling channels empty should neither overfill nor dry out. Since the solution to Equation (5.12) permits the calculation of the changes in the velocity and void fraction due to the varying heat load, the variation in the liquid mass flowrate may be easily found ($G_1 = \rho_2^0 v A_t (1 - \alpha_1)$). This flowrate is shown as a function of time and position in Figure 5.6a. A subset of this information is shown in Figure 5.6b. Notice that initially the liquid mass flowrate near the outlet of the channel ($L = 600$ m) exceeds the value closer to the inlet of the channel. However, at longer times the liquid mass flowrate at the outlet is less than near the inlet. This is a result of the void fraction increasing with time and distance along the channel.

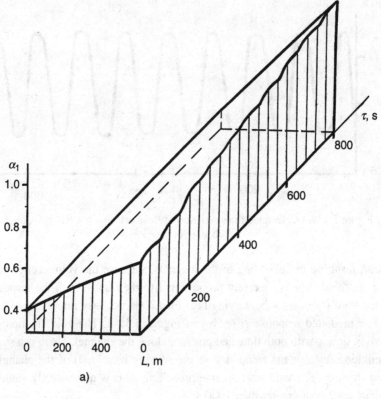

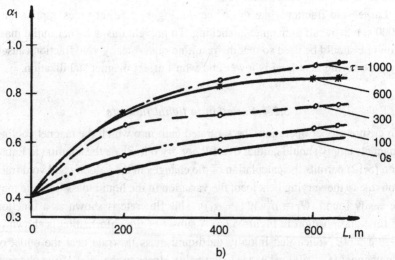

Figure 5.5. (a) Calculated change in void fraction as a function of time and position in the UNK two-phase channel due to a varying heat load. (b) Calculated void fraction as a function of position at several different times in the UNK two-phase cooling channels under varying heat load.

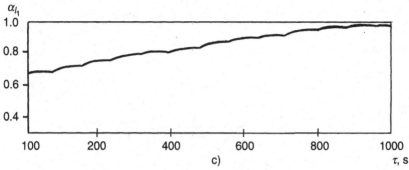

Figure 5.5. (c) Predicted void fraction change with time at the outlet of the UNK two-phase channel.

The liquid mass flowrate at the exit of the cooling channel (and thus entering the saturated bath) is shown as a function of time in Figure 5.6c. Region I on the plot shows the flowrate corresponding to the steady-state heat load. Region II shows the response to the varying heat load. There is at first an increase in the flowrate over the time period τ_1. Integrating the flowrate over this area will determine if the bath is in danger of overfilling. This integration yields a value of 0.4 to 0.5 m^3 of liquid helium. The bath in the UNK system has a volume of 1.96 m^3 and is initially filled to between 50 and 60 percent. The 0.5 m^3 additional liquid will only add another 26 percent to the level, so in these conditions the tank will not be overfilled. If the tank is in danger of overfilling during this response to the varying heat load, the flowrate must be decreased or the tank size changed to prevent this occurrence. Likewise over time interval τ_3 the liquid mass flowrate reaches its new quasi steady state. If the initial flowrate is too low, then the liquid flowrate at the outlet of the channel might be zero ($\alpha_1 = 1$) and the bath will dry out.

5.2.5 Summary of the model results

The solution to Equation (5.12) permits the determination of the flow parameters (e.g., velocity, void fraction, pressure) during the varying heat load as a function of both position and time. These parameters all show a similar behavior in that there is an initial transient after which the parameter approaches a quasi steady state. Superimposed on this change is an oscillation whose frequency is very close to that of the varying heat load that is driving the perturbation.

The knowledge gained from this analysis allows the designers to make important decisions on the required flowrate – to avoid superheating of the vapor or dry-out of the saturated bath – and on the size of the bath itself.

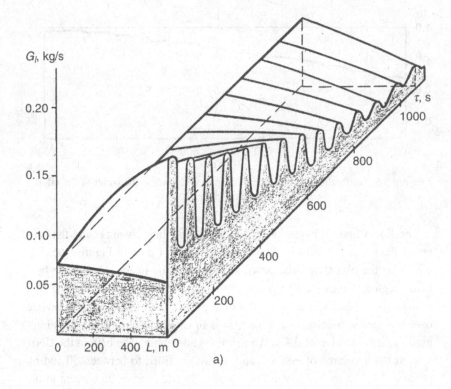

a)

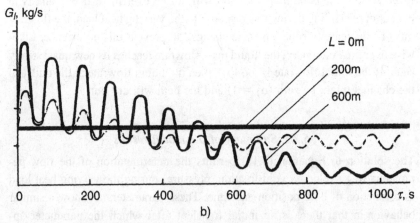

b)

Figure 5.6. (a) Calculated liquid mass flowrate as a function of time and position
in the UNK two-phase channel due to a varying heat load. (b) Calculated liquid
mass flowrate as a function of time in the UNK two-phase channel due to a
varying heat load.

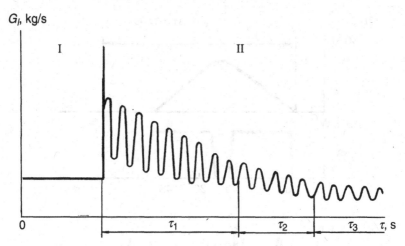

Figure 5.6. (c) Liquid mass flowrate at the UNK two-phase channel outlet as a function of time. I = steady-state heat load; II = varying heat load.

The length of time that it takes the two-phase flow to reach its new quasi steady state for a large ($L/D \gg 10^3$) system such as the UNK is seen to be tens of minutes. This time is important in the proper design of control systems and means that the numerical solution to Equation (5.13) must be carried out to at least this time. A large amount of computer memory is therefore needed to solve the problem.

5.3 Experimental studies

The first large accelerator to use flowing two-phase helium for magnet cooling was the Fermilab Tevatron. A numerical model was developed (Barton, Mulholland, & Nicholls 1981) to look at the response of the cryogenic system to the varying heat load. As in the case of the model developed for the UNK system, the Fermilab model predicts that the flow velocity will oscillate at near the same rate of the magnet ramping (see Figure 5.7a,b). A limited amount of data was taken to verify the model.

In order to check the model described in Section 5.2 a transient flowing two-phase helium experiment was built and operated (Ladohin & Gorbachev 1990). This experiment is shown schematically in Figure 5.8. In this test, helium flows from a liquid storage tank (whose pressure is maintained constant at $P_0 = 1.4 \times 10^5$ Pa) into a coiled test section with a 4.6 m long, 2.36 mm ID unheated section and a 13 m long, 5.4 mm ID uniformly heated section. By the end of the test section, all the helium has been converted to gas and the gas is collected

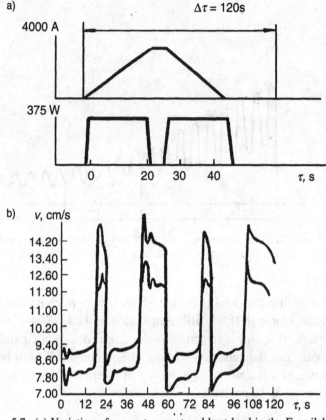

Figure 5.7. (a) Variation of magnet current and heat load in the Fermilab Tevatron accelerator. 1 cycle equals 120 s (Barton, Mulholland, & Nicholls 1981). (b) Predicted variation of flow velocity in the Tevatron magnets due to a varying heat load (Barton et al. 1981) (© 1981 IEEE).

by a large gas storage tank. During the experiment, the flowrate entering the gas storage system, the amount of heat applied, and the pressure differential across the heated section are measured as a function of time.

The calculated flow parameters for the experiment under a steady state heating of 2.5 W are shown in Figure 5.9. Because of wall friction the pressure in the test section drops as the fluid moves through the coil. The decreasing pressure results in a decrease in the saturation temperature of the two-phase flow. Also shown in the figure is the measured pressure drop across the heated section. The predicted and measured pressure drop agree to within 18 to 20 percent, with the measured value being consistently higher. This is thought to result from entry effects and other local resistances in the heated section that aren't well described in the model.

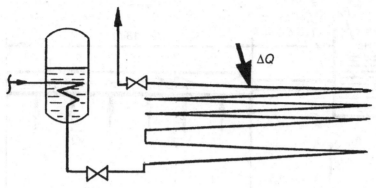

Figure 5.8. Schematic of transient two-phase flow experiment (Ladohin & Gorbachev 1990).

To examine the transient response of the system, a step increase is made to the heat applied to the test section. Figure 5.10 shows an example of such a change. Since this change in heat load is a step function rather than a periodic function, we would expect to see a transient change in the flow parameters followed by an approach to a new steady state. A quasi-steady state in which the flow parameters oscillate, like that seen in the modeling of the UNK system, would not be expected. The measured change in the pressure gradient across the heated section is shown in Figure 5.11 at two different variations in the heat load. Notice that the time in which the system approaches a new steady state is much shorter than that seen in the case of the UNK system (~10 s vs. greater than 1000 s). This is a result of the difference in channel length (13 m vs. ~650 m) between the two systems.

The amount of additional heat applied in the experiments is small enough so that the approximations made in the development of the model of Section 5.2 are valid. Adapting this model to the experimental configuration allows predictions to be made about the velocity and pressure distribution as a function of time and space as a result of the change in the heat load. Figure 5.12 a,b shows the model predictions for the case of a step increase of 20 W.

Because of the large size of the liquid supply tank its pressure is constant at 1.4×10^5 Pa during the change in heat load. For a sufficiently high change in heat load, however, the pressure in the two-phase channel may exceed that of the supply tank. Under these conditions, the two-phase helium is expelled from the channel both downstream into the gas reservoir and upstream into the liquid supply tank. As the pressure in the channel then falls again helium flows back from the storage tank into the channel. Figure 5.13 shows the calculated variation of peak channel pressure with the changes in applied heat

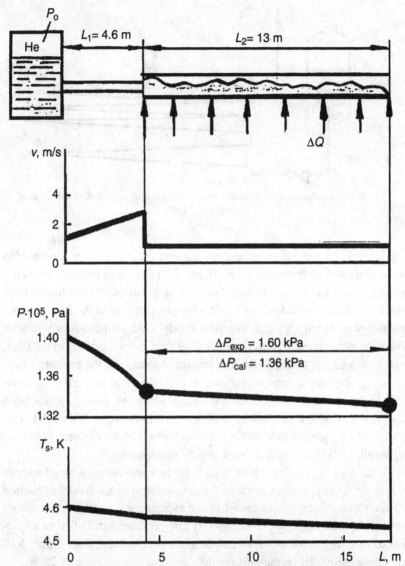

Figure 5.9. Steady-state flow parameters in the two-phase flow experiment
(Ladohin & Gorbachev 1990). ——— = model predictions; ● = experimental data.

load. Above a step increase of approximately 25 W the peak channel pressure
exceeds the tank pressure. The predicted pressure distribution for the case of
a 30 W increase is shown in Figure 5.14.

The comparison of predicted and measured pressure drops in the two-phase

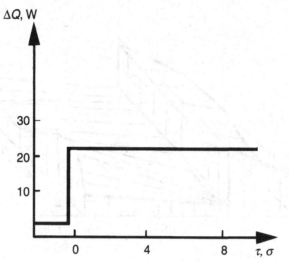

Figure 5.10. Description of step heat input into the transient two-phase experiment (Ladohin & Gorbachev 1990).

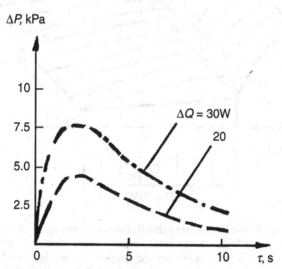

Figure 5.11. Measured change of pressure drop across the heated section of the transient two-phase experiment (Ladohin & Gorbachev 1990).

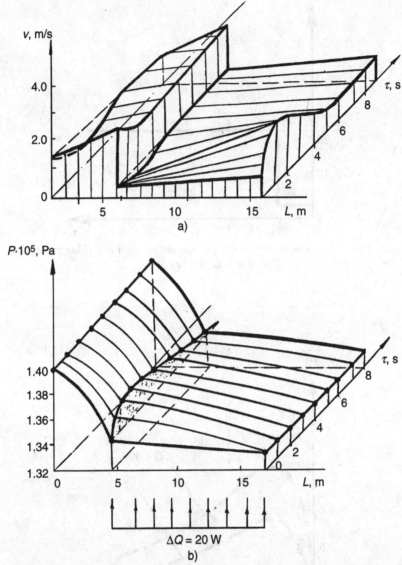

Figure 5.12. (a) Calculated velocity distribution as a function of time and position at a heat input of 20 W in the transient two-phase experiment. (b) Calculated pressure distribution as a function of time and position at a heat input of 20 W in the transient two-phase experiment (Ladohin & Gorbachev 1990).

channel is shown in Figure 5.15. The model results agree with the measured data to within 18 to 20 percent. The good agreement is a result of the approximations of the model (i.e., $v_1 \sim v_2 \sim v$, $T_1 \sim T_2 \sim T_s$, $dv/dt \sim dP/dt \sim 0$) being valid for the physical conditions of the experiment. A much more

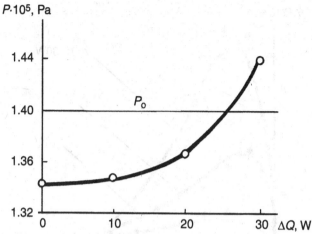

Figure 5.13. Predicted peak pressure in the heated channel of the transient two-phase experiment as a function of applied heat load (Ladohin & Gorbachev 1990).

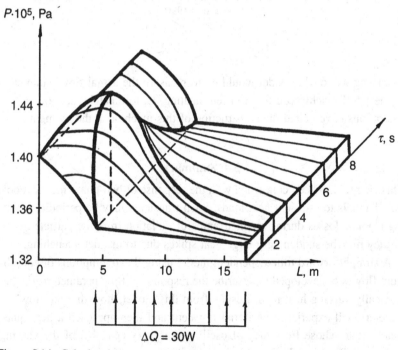

Figure 5.14. Calculated pressure distribution as a function of time and position at a heat input of 30 W in the transient two-phase experiment (Ladohin & Gorbachev 1990).

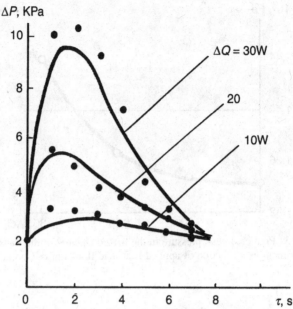

Figure 5.15. Comparison of calculated and measured pressure drops in the transient two-phase experiment as a function of time and heat input (Ladohin & Gorbachev 1990).

satisfying test of the model would be to measure the actual flow parameters in the UNK machine during transient heating. Unfortunately, economic considerations have halted the construction of this machine at the moment.

5.4 Summary

This chapter has been concerned with the response of two-phase magnet cooling channels to transient heat loads. Such transients may be periodic, resulting from ac losses during magnet ramps of radiation from the particle beam, or they may be sudden very large heat spikes due to magnet quenching.

A straightforward time-dependent model using the assumptions of equilibrium flow was developed to describe the response of flow parameters to a periodically varying heat load. Results from this model show that the flow parameters will experience an initial transient and then approach a new quasi steady state whose frequency of oscillation is very near that of the varying heat load. This model allows designers of the cryogenic systems cooling the magnets to make informed decisions on the sizing of flowrates, channel diameters, and reservoirs that ensure reliable operation of the superconducting magnets.

The model also predicts that the time for the cooling system to achieve a new quasi steady state is on the order of tens of minutes. This results from the size of these accelerator magnet systems and the typically slow speed of the two-phase flow. The time to a new quasi steady state is particularly important when considering the design of the control systems for these accelerators.

Operating experience from the Fermilab Tevatron accelerator (Barton et al. 1981) has qualitatively confirmed some of the results of the model. Quantitative agreement with model predictions was seen in a smaller laboratory experiment (Ladohin & Gorbachev 1990). Although the model was originally developed as part of the design of the Russian UNK machine, it could be applied with appropriate modifications to other magnet systems as long as the initial assumptions of the model remain valid.

Bibliography

Alexeyev, A. I., Filippov, Y. P., and Mamedov, I. S. 1991. "Flow Patterns of Two-Phase Helium in Horizontal Channels." *Cryogenics* 31:330–337.

Baker, O. 1954. "Simultaneous Flow of Oil and Gas." *Oil Gas J.* 53:185.

Bald, W. B., Zust, H., and Wooster, W. G. 1977. "An Apparatus for Visualizing Continuous Flow Boiling in Liquid Helium I." *Cryogenics* 17:33–42.

Bald, W. B., and Hands, B. A. 1974. "Cryogenic Heat Transfer Research at Oxford: Part II – Flow Boiling." *Cryogenics* 14:179–197.

Barton, H. R., Mulholland, G. T., and Nicholls, J. E. 1981. "Multi-level Computer Simulation of the Energy Saver Cooling Provisions." *IEEE Transactions on Nuclear Science* NS–28, 3:3306–3308.

Barton, H. R., Clausen, M., Horlitz, G., Knust, G., and Lierl, H. 1986. "The Refrigeration System for the Superconducting Proton Ring of the Electron Proton Collider HERA." *Adv. Cryo. Engr.* 31:635–645.

Bézaguet, A., Casa-Cubillos, J., LeBrun, Ph., Marquet, M., Tavian, L., and van Weelderen, R. 1994. "The Superfluid Helium Model Cryoloop for the CERN Large Hadron Collider (LHC)." *Adv. Cryo. Engr.* 39A:649–657.

Boom, R. W., El-Wakil, M., and McIntosh, G. E. 1978."Experimental Investigation of the Helium Two Phase Flow Pressure Drop Characteristics in Vertical Tubes." *Proc ICEC* 7:468–473.

Brédy, Ph., Neuvéglise, D., François, M. X., Meuris, C., and Duthil, R. 1994. "Test Facility for Helium I Two-Phase Flow Study." *Cryogenics* 34 (ICEC Supplement):361–364.

Brubaker, J. C., Fuerst J. D., and Norris, B. L. 1992. "Characterization of Two Phase Flow in a Fermilab Tevatron Satellite Refrigerator." *Adv. Cryo. Engr.* 37A:171–179.

Budrik, V. V. et al. 1990. "A Method of Void Fraction Calculation for Vapor or Gas–Liquid Flow Conditions" (in Russian). *Fizika Nizkih Temperatur* 16 (4):428–433.

Casas, J., Cyvoct, A., Lebrun, Ph., Marquet, M., Tavian, L., and van Weelderen, R. 1992. "Design Concept and First Experimental Validation of the Superfluid Helium System for the Large Hadron Collider (LHC)." *Cryogenics* 32 (ICEC Supplement):118–121.

Danilov, V. V., Filippov, Y. P., and Mamedov, I. S. 1990. "Peculiarities of Void Fraction Measurement Applied to Physical Installation Channels Cooled by Forced Helium Flow." *Adv. Cryo. Engr* 35:745–754.

Delhaye, J. M., Giot, M., and Riethmuller M. L. 1981. *Thermohydraulics of Two-Phase Systems for Industrial Design and Nuclear Engineering*. New York: Hemisphere.

Dyachuk, M. I. 1991. "Prediction of Pressure Fluctuations in Cryogenic Gasification Systems and Reduction to Nominal Values" (in Russian). Thesis for a Degree in Technical Sciences, Moscow.

Filin, N. V., and Bulanov, A. V. 1985. *Liquid Cryogenic Systems* (in Russian). Leningrad: Mashinostroeniye.

Filina, N. 1992. "Systems Ensuring Cryostability of Superconducting Magnets." *Cryogenics* 32 (ICEC Supplement):384–389.

Filippov, Y. P., and Alexeyev, A. I. 1994. "Improvement of Radiofrequency Void Fraction Sensors." *Adv, Cryo. Engr.* 39:1097–1104.

Filippov, Y. P., Alexeyev, A. I., Ivanov, A. V., and Lunkin, B. V. 1993. "Expanding Application of Cryogenic Void Fraction Sensors." *Cryogenics* 33:828–832.

Filippov, Y. P., Alexeyev, A. I., Lebedev, N. I., Mamedov, I. S., and Romanov, S. V. 1992. "Monitoring of Cryogenic Flows: Realization of the Radiofrequency Method." *Adv. Cryo. Engr* 37B:1461–1470.

Filippov, Y. P., Mamedov, I. S., and Selyunin, S. Y., 1988. "Characteristics of Horizontal Two-Phase Helium Flow at Low Mass Velocities." *Proc. ICEC* 12:198–201.

Gorbachev, S. P. 1987. "Modelling of the Working Processes of Cryostability Systems of Large Superconducting Devices" (in Russian). PhD dissertation, NPO Cryogenmash.

Gun, S. Z., Filippov, Y. P., and Zinchenko, S. I. 1985. "Comparison of Methods for Superconducting Dipole Magnet Cryostabilization" (in Russian). *Trudi Vsesoyuznogo Soveshchaniya pouskoriyelyam zarayajennih chastits*: 350–359.

Gun, S. Z., and Filippov, Y. P. 1984. "Calculation of Hydraulic Resistance in Channels Containing Two-Phase Helium Flows" (in Russian). *Teploenergetika* 3:19–23.

Hagedorn, D., Leroy, D., Dullenkopf, P., and Haas, W. 1986. "Monitor for the Quality Factor in Two-Phase Helium." *Adv, Cryo. Engr.* 31:1299–1307.

Hands, B. A. 1975. "Pressure Drop Instabilities in Cryogenic Fluids." *Adv. Cryo. Engr.* 20:355–369.

Haruyama, T. 1987. "Optical Method for Measurement of Quality and Flow Patterns in Helium Two-Phase Flow." *Cryogenics* 27:450–453.

Huang, X. 1994. "Hydrodynamic Study of One Dimensional Two-Phase Helium Flow." PhD thesis, University of Wisconsin–Madison.

Huang, X., and Van Sciver, S. W. 1994. "An Investigation into the Performance of a Venturi in Two-Phase Helium Flow." *Adv. Cryo. Engr.* 39:1065–1071.

Johnson, P. 1983. Safety Chapter in *Liquid Cryogens* Vol. 1. Ed. Williamson, K. D., and Edeskuty, F. J. Boca Raton: CRC Press.

Katheder, H., and Sußer, M. 1991. "Measurement Device with Cold Oscillator for Measuring Vapour Content in Helium Two-Phase Flow." *Cryogenics* 31:327–329.

Katheder, H., and Sußer, M. 1988. "Results of Flow Experiments with Two-Phase Helium for Cooling of Superconductors." *Proc ICEC* 12:207–211.

Katheder, H., and Sußer, M. 1986. "Pressure Drop in Adiabatic and Nonadiabatic Horizontal Two-Phase Helium Flow." *Proc ICEC* 11:429–433.

Keilin, V. E., Klimenko, E. Y., and Kovalev, I. A. 1969. "Device for Measuring Pressure Drop and Heat Transfer in Two-Phase Helium Flow." *Cryogenics* 19:36–38.

Kuzmenko, G. P. 1979. "The Investigation and Intensification of the Heat Exchange Process Under Cryogenic Liquid Film Boiling in Gasifiers" (in Russian). Thesis for Candidate's Degree in Technical Science, Moscow High Technical School.

Ladohin, S. D., and Gorbachev, S. P. 1990. "Transient Stratified Two-Phase Helium Flow in Horizontal Channels" (in Russian). *Himicheskoye i neftyanoe mashinostroenie* 4:15–17.

LeBrun, Ph. 1994. "Superfluid Helium Cryogenics for the CERN Large Hadron Collider Project at CERN." *Cryogenics* (ICEC Conference Supplement) 34:1–8.

Lockhart, R. W., and Martinelli, R. C. 1949. "Proposed Correlation of Data for Isothermal Two-Phase, Two-Component Flow." *Chem. Engr. Prog.* 45:39.

Mahe, M. J. 1991. "Study of Heat Transfer Properties of Two-Phase Helium in Forced Convection" (in French). PhD thesis, University of Pierre and Marie Curie, Paris, France.

Mandane, J. M., Gregory. G. A., and Aziz, K. 1974. "A Flow Pattern Map for Gas–Liquid Flow in Horizontal Pipes." *Int. J. Multiphase Flow* 1:537–553.

Mamedov, I. S. 1984. "Two-Phase Helium Flow Regimes in Horizontal Channels" (in Russian). Joint Institute of Nuclear Research–Dubna Report P8–84–156.

Nakagawa, S., Haraguchi, K., Kamiya, S., Iwata, A., Yoshiwa, M., Tada, E., Kato, T., Okuno, K., and Shimamoto, S. 1984. "Pressure Drop and Heat Transfer in Helium Two Phase Flow." *Proc ICEC* 10:570–573.

Neuvéglise, D., Brédy, Ph., François, M. X., and Meuris, C. 1994. "Heat Exchange in Horizontal Helium Two-Phase Flow." *Cryogenics* 34 (ICEC Supplement):357–360.

Nigmatulin, R. I. 1990. *Dynamics of Multiphase Media*, Vols. 1, 2. New York: Hemisphere.

Nigmatulin, R. I., Filina, N. N., Kroshilin, V. E., and Dyachuk, M. I. 1990. "Pressure Fluctuations in Gasification Systems" (in Russian). *Izvestiya Akademii Nauk ssr, Energetika i Transport*, 28(2): 134–138.

Rode, C., Brindza, P., Richied, D., and Stoy, S. 1980. "Energy Doubler Satellite Refrigerator Magnet Cooling System." *Adv, Cryo. Engr.* 25:326–334.

Richtmeyer, R., and Morton, K. W. 1967. *Difference Methods for Initial Value Problems*. New York: Interscience.

Sauvage-Boutar, E., Meuris, C., Poivilliers, J., and François, M. X. 1988. "Observation of Two-Phase Helium Flows in a Horizontal Pipe." *Adv. Cryo. Engr.* 33:441–447.

Schwerdtner, M. V., Stamm, G., Tsoi, A. N., and Schmidt, D. W. 1992. "The Boiling-up Process in He II: Optical Measurements and Visualization." *Cryogenics* 32:775–780.

Stirikovich, M. A., Polonsky, V. S., and Tsiklauri, G. V. 1982. *Heat and Mass Transfer and Hydrodynamics in Two Phase Flows at Atomic Power Stations* (in Russian). Moscow: Nauka.

Taitel, Y., and Dukler, A. E. 1976. "A Model for Predicting Flow Regime Transitions in Horizontal and Near Horizontal Gas–Liquid Flow." *AIChe Journal* 22:47–55.

Theilacker, J. C., and Rode, C. 1988. "An Investigation into Flow Regimes for Two-Phase Helium Flow." *Adv. Cryo. Engr.* 33:391–398.

Van Sciver, S. W. 1993. "Recent Advances in Helium Heat and Mass Transfer." *Recent Advances in Cryogenic Engineering* ASME HTD 267:1–11.

Van Sciver, S. W. 1986. *Helium Cryogenics*. New York: Plenum Press.

Vishniev, I. P., Migalinskaya, L. N., and Lebedeva, I. B. 1982. "Experimental Study of Two-Phase Helium Flow Hydraulic Resistance in Channels" (in Russian). *Injernero–fzicheskiyi* 43(2).

Zust, H. K., and Bald, W. B. 1981. "Experimental Observations of Flow Boiling of Liquid Helium I in Vertical Channels." *Cryogenics* 21:657–661.

Index